GEOTECHNICAL SPECIAL PUBLICATION NO. 110

COMPUTER SIMULATION OF EARTHQUAKE EFFECTS

PROCEEDINGS OF SESSIONS OF GEO-DENVER 2000

SPONSORED BY
The Geo-Institute of the American Society of Civil Engineers

August 5–8, 2000
Denver, Colorado

EDITED BY
Kandiah Arulanandan
A. Anandarajah
Xiang Song Li

American Society
of Civil Engineers
1801 ALEXANDER BELL DRIVE
RESTON, VIRGINIA 20191–4400

Abstract: This proceedings contains nine papers presented at the GeoDenver 2000 conference on Computer Simulation of Earthquake Effects. The papers cover non-destructive site characterization and property evaluation, constitutive modeling and numerical procedures and applications. There are two papers presenting details on the use of an electrical technique for evaluating in situ properties. The third paper presents the use of ground penetrating radar for predicting changes in soil density during soil liquefaction. Another paper presents details of how shear wave velocities measured in situ are capable of predicting the liquefaction potential. The fifth paper describes the constitutive modeling of flow liquefaction and cyclic mobility in detail. There are two papers dealing with the soil-pile structure interaction in liquefiable soils. The other two papers cover non-destructive electrical in situ site characterization to quantify the initial state parameters and constitutive model constants representative of the site for use in verified numerical procedures.

Library of Congress Cataloging-in-Publication Data

Geo-Denver 2000 (2000 : Denver, Colo.)
    Computer simulation of earthquake effects : proceedings of sessions of Geo-Denver 2000 / sponsored by The Geo-Institute of the American Society of Civil Engineers ; edited by Kandiah Arulanandan, A. Anandarajah, Xiang Song Li : August 5-8, 2000, Denver, Colorado.
        p. cm. – (Geotechnical special publication ; no. 110)
    Includes bibliographical references and index.
    ISBN 0-7844-0523-9
    1. Soil dynamics--Computer simulation--Congresses. 2. Earthquakes--Computer simulation--Congresses. 3. Soil liquefaction--Computer simulation--Congresses. 4. Soil structure interaction--Computer simulation--Congresses. I. Arulanandan, Kandiah. II. Anandarajah, A. III. Li, Xiang Song. IV. American Society of Civil Engineers. Geo-Institute. V. Title. VI. Series.

TA711.A1 G46 2000
624.1'5136—dc21   00-042141

Geotechnical Special Publications

Abstract

In recent years the sophistication of soil modeling has reached a degree at which it is possible to make calculations of the response of soil to dynamic input. The constitutive relations have been incorporated in computer codes and used to make assessments of ground responses such as acceleration, pore water pressure and deformations time histories. Most of the current methods treat the problem as equivalent to the calculated responses based on the total stress approach using laboratory cyclic triaxial and simple shear test data obtained on different samples existing at different locations at the site. This type of approach has been shown to be inadequate for the prediction of liquefaction-induced deformations and pore pressure response of soil deposits (Arulanandan and Scott 1993). Within the last two decades a small group of investigators have been conducting research on the representation of soil behavior by rigorous mathematical models of the non-linear constitutive behavior of soils and the use of non-destructive testing for site characterization. Within the last 5 years, such models have proved able to describe liquefaction induced deformation using centrifuge model tests to describe liquefaction induced deformation (Velacs Extension Project 1996 "Application of Numerical Procedures in Geotechnical Earthquake Engineering" National Science Foundation and Caltrans sponsored Workshop held at UC Davis). During the recent years user friendly numerical procedures such as SUMDES has also been developed and verified by using centrifuge models prepared in the laboratory. The application of VELACS concept, however to practice requires site characterization to obtain the initial state parameters (coefficient of hydraulic conductivity, coefficient of earth pressure at rest, porosity) and constitutive model constants (slope of the virgin compression line, slope of the rebound line, friction angle, maximum shear modulus, hardening parameters, position of critical state line in e-p space, etc.) representative of the site and verified numerical procedures using the results of the response of instrumented sites to earthquakes.

When all of the above has been put together, the power of computers will make it possible to carry out essentially deterministic calculations and evaluation of the response of the soil in an earthquake, given geophysical and soil input data representative of the site. The change in soil mechanics and soil mechanics thinking is analogous to the change that occurred in the early 1960's in structural engineering.

The objective of the Geotechnical Special Publication (GSP) is to present papers in the following three areas: 1. Non-destructive site characterization and properties; 2. Constitutive modeling and numerical procedures; 3. Application to dynamic site response and liquefaction-induced deformation of instrumented sites and soil-pile interaction in liquefiable sites subjected to large magnitude earthquakes to demonstrate numerical simulation of earthquake effects.

Preface

The worldwide total expenditure involved in infrastructure development associated with losses due to hazards such as earthquakes is estimated to be on the order of several billions of dollars. The characteristics and behavior of soil at any project location, either onshore or offshore, have a major influence on the success, economy, and safety of the work. Dealing with an earth material properly requires knowledge of the characteristics of the soil, the mechanisms causing soil behavior, and the properties that are necessary to analyze the influence of earthquake effects on soil behavior. In order to assess before the event the consequences due to any hazards, improvements in the current methods of soil and site characterization are necessary. Thus, for the *computer simulation* of earthquake effects of any infrastructure system, non-destructive characterization methods are required to obtain the initial state parameters, (porosity, coefficient of permeability, and coefficient of earth pressure at rest), and constitutive model constants such as compression index, swell index, friction angle, stress ratio required to cause liquefaction, hardening parameters, etc., that are representative of the site. In addition, a verified numerical procedure is necessary to evaluate the response of earthquake effects.

"Because of the difficulty of obtaining and testing soil samples, this area has not advanced as much as the geology and geophysical problems. Standard penetration and cone penetration tests do not supply enough information for calculation involving even simple material descriptions, far less the advanced material properties which are now employed in the computer models, and therefore other techniques are required. We have probably reached the limit of complexity in laboratory tests, as indicated by the fact that no new devices have appeared for many years, but further advances via in-place testing appear possible." (Dr. Ron F. Scott, Conference: Application of Numerical Procedures in Geotechnical Earthquake Engineering, Sponsored by NSF and CALTRANS, October 28-30, University of California, Davis, 1996)

In regard to the advancements in constitutive modeling and numerical methods, research over the past three decades has produced a variety of techniques. As a consequence, certain geotechnical problems, particularly those involving static problems, can be solved with reasonable degree of success. Techniques and concepts are still needed for modeling the behavior of earth structures and soil-structure interaction problems under dynamic loading. Among such problems, those involving soil liquefaction are particularly challenging. Difficulties are encountered not only in constitutive modeling, but also in numerical modeling. While it is difficult to include papers representing all facets of these challenging problems in this volume, representative papers are included. These papers present different approaches adopted for modeling the liquefaction behavior of sands, including a technique based on a two-surface concept and an associated flow rule, a technique based on a disturbed-state concept, and a technique based on the state parameter with bounding surface hypoplastic concept. While these techniques are quite different from each other, they all provide satisfactory means of mathematically capturing the stress-strain behavior of liquefying sands in dense and loose states. With the aid of these models to describe the stress-strain behavior, fully coupled finite element analyses of soil-structure problems are carried out and the results are presented in two of the papers. Excellent agreement between theoretical and numerical results attests to the potential for use of these approaches now for the design of complex soil-structure systems. Methods for continued verification of numerical procedures, and methods for determining the needed fixed and initial parameters in situ are of paramount importance in promoting rational approaches in design.

Various techniques for characterizing soils exist at the present time, each providing different types of information. Widely used techniques include SPT and CPT techniques. While these techniques have served the community reasonably well, new techniques such as the shear wave velocity techniques, electrical methods, ground penetrating radar, etc., are found to have potential for providing new, reliable information. Included in this volume are papers dealing with the use of these novel techniques for prediction of soil liquefaction during earthquakes. There are two papers presenting details on the use of an electrical technique for evaluating in situ constitutive model parameters that appear in advanced analysis techniques such as the fully coupled finite element techniques. One paper presents the details of how shear wave velocities measured in situ are capable of predicting the liquefaction potential. There is yet another paper on the use of ground penetrating radar for predicting changes in soil density during soil liquefaction. Results presented in these papers collectively demonstrate that these new techniques should be given serious consideration for predicting the liquefaction behavior of saturated sands during earthquakes.

Nondestructive in situ methods have been developed to accurately quantify soil structure, from which soil properties and model parameters can be obtained. The combined use of these non-destructive methods and verified numerical procedures provides the most promising means for simulating the effects of earthquake on infrastructure components.

The paper peer review was made possible with the valuable contribution of the following reviewers:

A. Anandarajah
Xiang Song Li
Richard D. Woods
Leonard R. Herrmann
Chandrakant Desai
Yannis F. Dafalias
Stein Sture
Raj Sidharthan
Kandiah Arulanandan
Kathiravetpillai (Siva) Sivathasan

Special thanks goes to Kathiravetpillai (Siva) Sivathasan for the continued support in the preparation of the Geotechnical Special Publication. We wish to thank Garret Broughton for assisting in preparation of the cover page.

Finally, we would like to express our grateful appreciation to the authors for their valuable contributions and very hard work toward the success of the session. We hope that the proceedings will be of significant value to researchers and practitioners.

Kandiah Arulanandan
University of California

A. Anandarajah
The Johns Hopkins University

Xiang Song Li
The Hong Kong University of Science and Technology

Contents

DIELECTRIC DISPERSION METHOD FOR NON-DESTRUCTIVE QUANTIFICATION OF SOIL COMPOSITION

K. Arulanandan, M.ASCE[1], and C. Yogachandran M.ASCE[2]

ABSTRACT

The Maxwell-Wagner-Fricke theory for electrical conduction through heterogeneous media has been extended to develop a theoretical relationship for fluid-saturated porous media, accounting for particle orientation, porosity and the dielectric constants of the pore fluid, solid phase and mixture. This theory is further extended to account for the dielectric dispersion behavior of clay-water-electrolyte systems by considering the surface conductance of particles. The variation of the dielectric constant, ε, and the conductivity, σ, as a function of frequency is the electrical dispersion behavior. The magnitude of the variation of ε over the radio frequency range (1-100 MHz) is referred to as $\Delta\varepsilon_0$. Dielectric dispersion behavior is shown to be controlled by the mineral-solution interface characteristics as influenced by mineralogy and the pore fluid composition. $\Delta\varepsilon_0$ is shown to be related to swell potential and compression index. The electrical dispersion method presented, provides a rapid non-destructive method of quantifying the composition of soils.

KEYWORDS

Dielectric constant, dielectric dispersion, mineral composition, mineral solution interface characteristics, swell potential, compression index, surface conductance, soil classification

INTRODUCTION

In geotechnical engineering, soil classification is used for soil identification, estimation of properties, and translation of experience. By 1908, Atterberg had developed a classification of size fractions based on decimal multiples of 2μ and 6μ, with "clay fraction" defined as the percentage by weight of particles smaller than 2μ. However he realized that particle size alone provided an insufficient basis for the classification of cohesive soils, and decided that a measure of their "plasticity" in terms of liquid and plasticity limits provided to a large extent the additional criteria required. With some modification in nomenclature, this system has been used in geotechnical engineering for the

[1]Professor, Dept. of Civil Engineering, Univ. of California, Davis, CA 95616
[2]Geotechnical Manager, CH2MHill, Santa Anna, CA 92707

past 80 years.

Difficulties with the existing systems of soil classification, as well as the considerable variability between these systems, led to a critical review by Casagrande (1948) and the proposal of the Unified Soil Classification system, adopted in 1952 by the U.S. Corps of Engineers and Bureau of Reclamation, and subsequently by many other organizations (USBR 1963). The Unified Soil Classification System, like all other procedures based on grain size and the properties of remolded materials, can not always reflect the compositional characteristics of a soil, nor can it account for soil structure, state, ambient conditions, or other factors which determine the specific values of the various soil properties. Additionally, the commonly used boundary between particle sizes of non-clay minerals and clay minerals of 2 μm may lead to incorrect classifications because, in natural soils, there are ordinarily particles of clay minerals that are larger than 2 μm in size as well as non-clay particles smaller than this size. Therefore, an arbitrary division based on grain size serves only as a rough measure in determining soil composition.

In a series of five papers, Lambe and Martin (1953-1957) reported compositional data for a large number of soils, then discussed the relationship between composition and engineering properties. These studies showed that the plasticity characteristics of soils containing mixtures of clay minerals such as found in natural soils may be less than would be predicted using knowledge of the properties of the pure clay minerals present and the percentage of each of the clay minerals in the soil as shown by the data in Figure 1 (Lambe & Martin, 1957). In addition, the percentage of clay mineral by weight, as determined by x-ray diffraction or differential thermal analysis, can be more or less than the percentage clay size (2 μm), as shown by the data in Figure 2, (Lambe and Martin, 1957, and Basu and Arulanandan, 1973). Aggregation, cementation and inter stratification of clay minerals are usually cited as the reasons for clay mineral contents in excess of the clay size content.

Atterberg Limits are often unable to properly estimate soil properties. For two soil samples with the same gradation, both containing a small percentage of fines (e.g. 5%), it would not be possible to determine Atterberg Limits for either sample. However, if the fines in one sample contained clay minerals, and the fines in the other contained non-clay minerals such as silica flour, the mechanical properties of the two soils would be different. Lambe and Martin found that Atterberg Limits could not always accurately evaluate compressibility. For soils indigenous to St. Louis and elsewhere they found that "very large changes in volume can occur with changes in load or changes in moisture conditions and that these changes are partially reversible. The Atterberg Limits of the St. Louis soil (Sample 19, [L.L.] = 58, [P.L.] = 22) do not suggest such unusual compressibility characteristics" (Lambe and Martin, 1957).

The parameters used to identify swell potential, activity and percent finer than 2µm, are not adequate for accurate evaluation of percent swell for natural soils. Seed, et al., (1962) correlated activity to percent finer than 2 µm to evaluate the swell potential of soils. Their evaluations were reasonably accurate for the artificial soils they tested. A study of both natural and artificial soils made by Basu and Arulanandan (1973) evaluated the ability of the activity index and the percent clay size to evaluatre swell potential. The composition, percent clay size, activity and percent swell for each sample is tabulated in Table 1. The percent swell for the samples was then plotted on the chart developed by Seed, et al., (1962), Figure 3. It can be seen that the percentage total swell predicted for samples 1PB, 3SC and Marysville Red (MR) is significantly lower than the measured value. Thus, while the predicted value is reasonable for artificial soils, the predictions are inconsistent for natural soils. The soil samples from Table 1 are also plotted in Skempton's (1953) activity chart, Figure 4. The three solid lines represent the three clay minerals montmorillonite, illite and kaolinite, from highest to lowest activity. These lines were developed by mixing quartz sand with varying percentages of the three clay minerals. Comparison of the samples to the solid lines shows that activity does not consistently follow mineralogy. For example, Soil 15 with 15% montmorillonite, is expected to plot near the solid line representing montmorillonite with an activity of 7.2. Soil 15 actually plots below the line representing kaolinite (Activity = 0.38), suggesting an entirely different mineralogy. Note that samples 1PB, 3SC, 11, 9, 3 and 8 show similar deviations. Thus activity is not an accurate predictor of soil mineralogy.

In the construction of a section of road in Marysville, California, the subgrade consisted of a silty soil with the following properties: L.L. = 46, P.I. = 11, percent less than 2μ = 20, percent less than 1μ = 14, percent sand size = 10, percent silt size = 70, and percent clay size = 20 (Arulanandan, 1974). Three different methods were used to evaluate the swell potential: Skempton's Activity (1953), Holtz and Gibbs percent of free swell (1956), and a chart developed by Seed et al. (1962). All of these methods predicted that the soil would have a low swell potential. However, during construction, the desired density could not be obtained regardless of the compactive effort due to the expansive nature of the soil. In order to determine the actual swell potential of the soil, free swell tests were conducted in the laboratory. Numerous samples of the soil were compacted then subjected to swell at various water contents and dry densities, in half inch deep by four inch diameter Teflon lined rings with porous plugs. The free swell of the samples varied up to 30% within the small range of water contents. This suggested that the soil should instead be considered to have a high swell potential (Arulanandan, 1974).

The previous examples have made several points about some of the existing methods used in conventional geotechnical analysis: (1) gradation analysis alone is insufficient to characterize soils unless the mineralogy of the fines is taken into consideration, (2) Atterberg Limits alone are insufficient for evaluating the volume change characteristics of some natural soils (Lambe and Martin, 1957), and (3) the percent less than 2μ and the activity criteria are also insufficient for evaluating swell potential of some natural soils.

Mineralogy is the primary factor controlling the sizes, shapes, and surface characteristics of the particles in a soil. The type of minerals and the interactions between the solid and fluid phases determine the swelling, compression, strength, and permeability of soils. Thus mineralogy and pore fluid composition are fundamental factors controlling geotechnical properties.

Arulanandan et. al, 1973, proposed a non-destructive method to characterize the type and amount of clay minerals and the mineral-solution-interface characteristics (electrical double layer) using electrical methods. The objective of this paper is to demonstrate the significance of the magnitude of dielectric dispersion in the radio frequency range for the non-destructive characterization of soil composition.

RADIO FREQUENCY ELECTRICAL DISPERSION AND SOIL COMPOSITION

The application of an electrical impulse in the form of a voltage to any geological material in the radiofrequency range produces a response which can be measured in terms of the dielectric constant ε and conductivity σ. In non-clay minerals, the ε and σ response is independent of frequency as demonstrated in Figure 5, whereas in hydrated soils with clay minerals, the ε and σ vary as a function of frequency as also shown in Figure 5. Figure 6 shows the variation of the dielectric and conductivity behavior (electrical dispersions) of saturated Marysville soil which contains clay minerals. The variation of ε or σ with frequency is called the electrical dispersion. The magnitude of dielectric dispersion, $\Delta\varepsilon_0$, which is denoted in Figure 5 and also shown for the Marysville soil in Figure 6, is defined as the change in ε measured over the radio-frequency range. The magnitude of dielectric dispersion is determined by using the high and low frequencies at which the dielectric constants in the dispersion curve level off. The magnitude of dielectric dispersion of saturated soils has been shown to be a function of the mineralogy and the mineral-solution interface characteristics (due to double layer polarization). Therefore, the magnitude of dielectric dispersion can be used to quantify the composition of soils. Arulanandan et. al (1973) have proposed the dielectric dispersion method to characterize the type and amount of clay minerals in soils.

It can be seen from the dielectric dispersion behavior of most natural soils as shown in Figure 7, (Sargunam, 1973), that if the dielectric measurements are made at 2 MHz and 50 MHz, the magnitude of dielectric dispersion, $\Delta\varepsilon_0$ can be obtained. For soils which have dielectric dispersions which extend beyond the range of 2 and 50 MHz, such as the Marysville soil, $\Delta\varepsilon_0$ may be extrapolated from the experimental data using a Cole-Cole plot (Arulanandan, 1966) as was done in Figure 6 for the Marysville soil. A theoretical basis for this new index, $\Delta\varepsilon_0$, and its use to characterize the mineralogy and the mineral-solution interface characteristics as influenced by the pore fluid composition and particle shape is presented herein. The applicability of the dielectric dispersion behavior of saturated soils for the evaluation of swell potential, and compression index is also examined.

ELECTRICAL CONDUCTION THROUGH HETEROGENEOUS MEDIA

Arulanandan, et al., (1985) extended Maxwell (1881) and Fricke's (1924, 1953) analysis of the electrical properties of suspended ellipsoidal particles in an alternating current field to provide a theoretical basis to describe the electrical dispersion behavior of transversely isotropic soils, and then applied the theory to the evaluation of porosity (Arulanandan, et al., 1985; Arulanandan, 1991). The theory developed to describe the electrical dispersion behavior is summarized below.

Maxwell (1881) derived the following expression for the conductivity of a heterogeneous media consisting of spherical particles in a dilute electrolyte suspension, assuming that the electric field of one particle does not influence the electric field of another:

$$\frac{k}{k_1} = \frac{2k_1 + k_2 - 2(1-n)(k_1 - k_2)}{2k_1 + k_2 + (1-n)(k_1 - k_2)}$$

where k is the conductivity of the medium, k_1 is the conductivity of the solution, k_2 is the conductivity of the particle, and n is the porosity. Wagner (1914) extended Maxwell's equation to consider the relaxation mechanisms of the dielectric constant and conductivity that occurs with heterogeneous media under alternating current which is generally referred to as the Maxwell-Wagner relaxation mechanism. In clay soils, charges accumulate at the interface (mineral solution interface, the electrical double layer) between the clay particle and the surrounding solution. Since this build-up of charges takes time, as the frequency is increased there will be less time for the charges to accumulate at the interface, which in turn decreases the system's ability to store electrical potential energy, and thus decreases the dielectric constant. When the frequency reaches a certain value, there will not be enough time for charges to accumulate at the interface, and at this point, the dielectric constant becomes independent of frequency. The value of the dielectric constant at this leveling-off frequency is defined as ε_∞.

Fricke (1953) considered the particles to be ellipsoid, with all particles oriented in one direction only in dilute suspensions. Results of the theoretical dielectric dispersions considered in this paper show that Fricke's theory is valid for higher volume concentrations. Dafalias and Arulanandan (1979, 1983) considered ellipsoid particles of any axial-ratio oriented in multiple directions based on a probability distribution of orientations.

Consider the soil particles to be spheroids with semi-axes a, b and c, as in Figure 8, where b equals c and b and c are not equal to "a" in general. When $b/a \geq 1$, the particles are modeled as oblate spheroids and when $b/a < 1$, the particles are modeled as prolate spheroids. The following analysis holds for both oblate and prolate spheroids.

Representing the multiple orientation of particles by appropriate probability density functions, as proposed by Dafalias and Arulanandan (1979, 1983), expressions were obtained for the vertical (F_v) and horizontal (F_h) formation factors as a function of porosity, n, as follows (Dafalias and Arulanandan, 1979, 1983)

$$F_v = \frac{k_1 - k_2}{k_v - k_2} = 1 + \frac{1-n}{n} f_v \tag{1}$$

$$F_h = \frac{k_1 - k_2}{k_h - k_2} = 1 + \frac{1-n}{n} f_h \tag{2}$$

where

$$f_v = \frac{P_\theta}{1 + \frac{k_2 - k_1}{k_1} A_a} + \frac{1 - P_\theta}{1 + \frac{k_2 - k_1}{k_1} A_b} \tag{3}$$

and

$$f_h = \frac{1}{2}\left[\frac{1 - P_\theta}{1 + \frac{k_2 - k_1}{k_1} A_a} + \frac{1 + P_\theta}{1 + \frac{k_2 - k_1}{k_1} A_b}\right] \tag{4}$$

In the above expressions k_1, k_2 are the complex electrical conductivities of the solution and the particle, respectively. k_v and k_h are the complex electrical conductivities of the composite medium in the vertical direction and the horizontal direction, respectively. Complex conductivity k is defined as

$$k = \sigma + j\omega\varepsilon \tag{5}$$

where σ is the conductivity, ε is the apparent dielectric constant, ω is the angular frequency and $j = \sqrt{-1}$. ω is related to the frequency, f, by the expression $\omega = 2\pi f$. The shape indices, A_b and A_a for a spheroid are defined as (Fricke, 1953):

$$A_b = \frac{1}{2(1 - R^2)}\left[\frac{R^2}{2\sqrt{1 - R^2}}\ln\left(\frac{1 - \sqrt{1 - R^2}}{1 + \sqrt{1 - R^2}}\right) + 1\right], \qquad \text{for } 0 \le R \le 1 \quad (6a)$$

and

$$A_a = \frac{1}{2\sqrt{R^2 - 1}}\left[\frac{R^2}{\sqrt{R^2 - 1}}\tan^{-1}\sqrt{R^2 - 1} - 1\right], \qquad \text{for } R > 1 \quad (6b)$$

and $A_a = 1 - 2 A_b$, where axial ratio, $R = b/a$ (Fricke, 1953). Some values of shape indices for special cases are given: for spherical particles $R = 1$ and $A_b = 1/3$, for laminae shaped particles $R \longrightarrow \infty$ and $A_b \longrightarrow 0$, and for long infinite cylinders $R = 0$ and $A_b = 1/2$. The orientation factor, P_θ, in eqs. (3) and (4) represents the multiple orientations of particles and is defined as

$$P_\theta = \int_0^{\pi/2} p(\theta)\cos^2\theta \, d\theta \tag{7}$$

where the probability density function, $p(\theta)$, characterizes the distribution of the orientation of semi axis "a" with respect to the vertical direction for $0 \le \theta \le \pi/2$ such that (Dafalias and Arulanandan, 1979)

$$\int_0^{\pi/2} p(\theta) \, d\theta = 1 \tag{8}$$

An explicit expression for k_v can be obtained by rearranging eq. (1) as

$$k_v = k_1 + (k_2 - k_1) \frac{(1-n)f_v}{n+(1-n)f_v} \tag{9a}$$

A similar expression for k_h can also be obtained by rearranging eq. (2) as

$$k_h = k_1 + (k_2 - k_1) \frac{(1-n)f_h}{n+(1-n)f_h} \tag{9b}$$

Equations (9a and 9b) give expressions for complex electrical conductivities in the vertical and horizontal directions k_v and k_h, respectively, in terms of the complex conductivities of the solution, k_1, the particles, k_2, porosity, n, the particle orientation, P_θ, and the axial ratio, R. When each of the complex conductivities in eqns. (9a and 9b) are replaced by the conductivities and dielectric constants as given in eq. (5) the following equations are obtained:

$$\sigma_v + j\omega\varepsilon_v = \sigma_1 + j\omega\varepsilon_1 + \frac{[\sigma_2 - \sigma_1 + j\omega(\varepsilon_2 - \varepsilon_1)](1-n)f_v}{n+(1-n)f_v} \tag{10a}$$

$$\sigma_h + j\omega\varepsilon_h = \sigma_1 + j\omega\varepsilon_1 + \frac{[\sigma_2 - \sigma_1 + j\omega(\varepsilon_2 - \varepsilon_1)](1-n)f_h}{n+(1-n)f_h} \tag{10b}$$

where:

$$f_v = \frac{P_\theta}{1 + \dfrac{\sigma_2 - \sigma_1 + j\omega(\varepsilon_2 - \varepsilon_1)}{\sigma_1 + j\omega\varepsilon_1} A_a} + \frac{1 + P_\theta}{1 + \dfrac{\sigma_2 - \sigma_1 + j\omega(\varepsilon_2 - \varepsilon_1)}{\sigma_1 + j\omega\varepsilon_1} A_b} \tag{11a}$$

$$f_h = \frac{1}{2}\left\{ \frac{1 - P_\theta}{1 + \dfrac{\sigma_2 - \sigma_1 + j\omega(\varepsilon_2 - \varepsilon_1)}{\sigma_1 + j\omega\varepsilon_1} A_a} + \frac{1 + P_\theta}{1 + \dfrac{\sigma_2 - \sigma_1 + j\omega(\varepsilon_2 - \varepsilon_1)}{\sigma_1 + j\omega\varepsilon_1} A_b} \right\} \tag{11b}$$

σ_v and σ_h are the conductivities of the system in the vertical and horizontal directions respectively, σ_1 and σ_2 are the conductivities of the solution and the particles respectively, and ε_1 and ε_2 are the dielectric constants of the solution and particles respectively.

Thus the real part of eqns. (10a and 10b) will give expressions for the variation of vertical conductivity and horizontal conductivities, σ_v and σ_h , as a function of ω, σ_1, ε_1, σ_2, ε_2, P_θ, R, and porosity n. The imaginary part of eqns. (10a and 10b) will give expressions for the variation of the vertical and horizontal dielectric constants of the system, ε_v and ε_h , as a function of ω, σ_1, ε_1, σ_2, ε_2, P_θ, R, and n. The dielectric constant, ε_1, is approximately equal to 80. The measured value of the dielectric constant of dry particles, ε_2, is about 4.5 (Arulanandan, 1973).

CONDUCTIVITY OF THE SOIL PARTICLE σ_2

The electrical conductance of a soil particle σ_2 depends on the mineralogy and the adsorbed pore fluid. The electrical conductivity of the particle may be defined as the sum of two conductivities. The first is due to the conductivity of the charged carriers of the particle (conductivity of the spheroid) and the second conductivity is due to the existence of the double layer in the interface between the clay particle and the surrounding pore fluid (surface conductance). For dry particles the double layer is not present and the conductivity of the particle is due only to the charges of the particle. For saturated systems, the double layer is formed and the particle conductivity is due to the conductivity of the particle and the conductance due to the double layer (surface conductance).

Weiler and Chaussiddon (1968) investigated the surface conductivity of spherical and ellipsoidal particles based on Okonski's (1960) theoretical approach. For the ellipsoidal particle, Weiler and Chaussidon indicated that the effective conductivity of the charged particle consists of two components, one arising from the surface conductance (the double layer), and the other from the conductivity of the ellipsoid which depends on the axis under consideration (Weiler and Chaussidon, 1968). Weiler and Chaussidon stated that in highly hydrated systems, such as saturated clay minerals, it is reasonable to consider the conductivity of the ellipsoid (particle) to be negligible when compared to the surface conductivity of the particle.

Clay minerals are thin laminae shaped particles. Oblate spheroids with R>>1 would appropriately model these laminae shaped particles. An extreme of these oblate spheroids are discs and plates. Surface conductance, (λ_s in mhos), of such an oblate spheroid (R>>1) is given by Weiler and Chaussidan (1968) as

$$\lambda_s = \frac{\pi a \sigma_s}{4}$$
(12)

where σ_s is the surface conductivity of the oblate spheroid in mho/cm due to the double layer. The specific surface area, SSA, (in cm^2/g) is defined as the surface area per unit weight of a circular disc (an extreme case of an oblate spheroid) and can be obtained as

$$SSA = \frac{2\pi b^2 + 2\pi ba}{\pi b^2 a \rho_s}$$
(13)

where ρ_s is the density of the particle, "b" is the radius of the disc, and "a" is the thickness of the disc. Equation (13) can be rewritten as

$$SSA = \frac{2}{\rho_s a} \left\{ 1 + \frac{1}{R} \right\}$$
(14)

Using equation (12) and (14) and eliminating "a" from both equations an expression for λ_s is obtained as (Yogachandran, 1988):

$$\lambda_S = \frac{\pi}{2(SSA)\rho_s}\left\{1+\frac{1}{R}\right\}\sigma_S \tag{15}$$

The surface conductance of the clay mineral can be estimated using equation (12). Since the conductivity due to the charges within the particle can be neglected, the conductivity of the particle, σ_2, will be approximately equal to σ_S.

EXPERIMENTAL RESULTS

The results of previous experimental investigations performed to determine the sensitivity of electrical dispersion characteristics to variations in clay type and amount, particle orientation, and pore fluid concentration are presented below (Arulanandan and Smith, 1973; Arulanandan, Basu and Scharlin, 1973; Alizadeh 1975). The experimental measurements of the conductivity and dielectric constant of soils at various frequencies were made using the Boonton 250 RX meter (Boonton Radio Corporation).

Influence of Clay Content on Dielectric Dispersion

The influence of clay content on dielectric dispersion behavior was examined by measuring the dielectric dispersion behavior of soils consisting of 20%, 30%, and 40% montmorillonite mixed with sand (Alizadeh, 1975). The results are presented in Figure 9 and the magnitude of dielectric dispersions, $\Delta\varepsilon_o$, of these soils are tabulated in Table 2. The $\Delta\varepsilon_o$ values are based on extrapolation of the theoretically optimized data shown as dashed lines. The results presented in Figure 9 and Table 2 show that the magnitude of dielectric dispersion increases with increasing clay content. Plots of the magnitude of dielectric dispersion vs. various clay content for kaolinite and illite as well as montmorillonite are shown in Figure 10 (Alizadeh, 1975).

Influence of Clay Type on Dielectric Dispersion

The influence of clay mineral type on the dielectric dispersion behavior was examined by measuring the dielectric constant as a function of frequency for a 40%

montmorillonite-sand mixture, a 66% illite-sand mixture, 100% kaolinite, and 100% silt size Silica flour. The measured dielectric constants and conductivities as a function of frequency for the above soils are shown in Figures 11a and 11b respectively (Arulanandan, Basu and Scharlin, 1973). The magnitude of dielectric dispersion, $\Delta\varepsilon_0$, of the soils shown in Figure 11a are presented in Table 2. The results presented in Figure 11a and Table 2 show that $\Delta\varepsilon_0$ depends highly on mineralogy and increases from smallest to largest in the order kaolinite, illite-sand mixture, and montmorillonite-sand mixture. Through Figure 11a it is shown that among the different clay minerals shown, montmorillonite exhibits the highest $\Delta\varepsilon_0$, followed by illite and then kaolinite which in its pure state, has a smaller $\Delta\varepsilon_0$ than the mixtures of montmorillonite with non-clay soil and illite with non-clay soil.

Influence of Concentration of Pore Fluid on Dielectric Dispersion

The magnitudes of dielectric dispersion of montmorillonite, illite, and kaolinite mixed with sand were measured at various sodium chloride electrolyte concentrations. The variation of the magnitude of dielectric dispersion as a function of percent clay fraction, and the electrolyte concentrations is shown in Figure 12 (Gu and Arulanandan, 1980). It can be seen that the magnitude of dielectric dispersion is lower for higher pore fluid concentrations. One possible explanation is that an increase in electrolyte concentration decreases the thickness of the double layer (the mineral-solution interface) and thus the charges are more strongly associated with the surface of the clay particle. This reduces the potential for polarization, thereby reducing the magnitude of dielectric dispersion.

EVALUATION OF PARAMETERS σ_1, σ_2, R, P_θ and n

The theoretical expressions given in equations (10a) and (10b) can be fitted to the experimental electrical dispersion data of soils by computer optimization of the parameters (σ_1, σ_2, R, P_θ and n) (Yogachandran, 1988). Since the variables in expressions (10a) and (10b) are complex numbers, the computer program was written using a complex number algorithm to separate the real and imaginary parts of these expressions. The real parts of these expressions yields the conductivities, σ_v and σ_h, and the imaginary parts yield the dielectric constants, ε_v and ε_h, as a function of frequency. The value of the surface conductance, λ_s, can then be calculated from the value of σ_2 necessary to fit the experimental results using equation (12) and assuming $\sigma_s \approx \sigma_2$.

The dielectric dispersions shown in Figures 9 and 11a are evaluated using expressions (10a) and (10b). The parameters required for the theoretical evaluations are σ_1, σ_2, R, P_θ, and n. The porosities of the soils are known. The axial ratios of pure montmorillonite, kaolinite, and sand are about 100, 10, and 1 respectively (Lambe and Whitman, 1969). The particle shape of illite is between the shape of montmorillonite and kaolinite. Therefore a value of 30 was used as the axial ratio for illite in this paper. The axial ratios of the mixed soils are estimated based on the percentage of clay mineral present in the mixed soils. Since the values of σ_1 and σ_2 are not known for these soils, appropriate values of σ_1 and σ_2 were chosen so as to fit the observed experimental dielectric and conductivity dispersions. The values of P_θ were also optimized so as to best fit the observed experimental dielectric and conductivity dispersions The theoretical dielectric dispersions are compared with the experimental results and are presented in Figures 9, and 11a. The conductivity dispersions for soils in Figure 11a were also evaluated and the results are shown in Figure 11b. The values of σ_1, σ_2, R, P_θ, and n used to theoretically evaluate the experimental results shown in Figures 9, 11a and 11b are presented in Table 2.

These theoretical optimizations shown in Figures 9, 11a and 11b are done mainly to demonstrate that the fundamental mechanism behind the dielectric and conductivity dispersions is indeed the Maxwell-Wagner-Fricke electrical dispersions. The optimizations show that reasonable values for R, P_θ, σ_1, σ_2 and known porosity, n, produce dielectric and conductivity dispersion curves which match the experimental data reasonably well.

The measured values of specific surface area of montmorillonite, illite, kaolinite and assumed value of SSA for sand are 600 m^2/g, 150 m^2/g, 20 m^2/g and 1 m^2/g respectively (Meegoda, 1985). Using a simple mixing rule, the specific surface areas of the mixtures of soils are calculated and are presented in Table 2. The calculated surface conductance, λ_s, values using equation 15 for the soils are also given in Table 2. The values of λ_s are on the order of 10^{-9} mho. These values of λ_s are in the range of values reported by others (e.g., Cremers, et al., 1966; Weiler and Chaussidan, 1968; Schwan et al., 1962; Fricke and Curtis, 1937).

It is evident from the above analysis and demonstrated by the results presented in Table 2 that the dielectric dispersion increases with an increase in the value of σ_2 (which is approximately equal to σ_S in hydrated systems) and that no dielectric dispersion is observed when $\sigma_2 = 0$. Further, the results show that $\Delta\varepsilon_0$ is influenced by clay type and amount, and the pore fluid composition. The dielectric dispersion of soils provides a basis for quantifying the composition of soil.

DIELECTRIC METHOD OF SOIL CHARACTERIZATION

Results presented in Figures 9, 11a and 11b show that clay minerals exhibit dielectric and conductivity dispersion behavior, i.e., variations of dielectric constant and conductivity with frequency. It is also demonstrated in Figures 11a and 11b that non-clay minerals (i.e. silica flour) exhibit no dielectric and conductivity dispersion behavior, that is the dielectric constant (ε) and the conductivity (σ) are independent of frequency. The magnitude and the manner in which values of ε and σ vary for soils as a function of frequency were shown to be due to surface conductivity σ_S (and to λ_s through equation 12) which largely influences the conductivity of the particle σ_2. The dielectric dispersion behavior is shown to increase with increasing values of σ_2. This value of σ_2 depends on clay mineralogy and the clay mineral-solution interface characteristics. Further, it was shown that if $\sigma_2 \approx 0$, (e.g., silica flour), then no variation of dielectric constant with frequency was observed.

Based on the above description, it is possible to classify all soils as belonging to two groups, clay minerals and non-clay minerals, with the former exhibiting dielectric dispersion behavior, the magnitude of which depends on amount and type of clay mineral and the mineral-solution interface characteristics, and the latter exhibiting no dielectric dispersion behavior. The magnitude of dielectric dispersion, $\Delta\varepsilon_0$, is a function of clay mineral type and amount and is influenced by pore fluid composition as is evident from the results shown in Table 2 and Figures 9, 11, and 12.

MAGNITUDE OF DIELECTRIC DISPERSION AND COMPRESSION INDEX

It has been shown that $\Delta\varepsilon_0$ is significantly influenced by the type and amount of clay mineral. The values of $\Delta\varepsilon_0$ are shown to increase in the sequence kaolinite < illite < montmorillonite. The compression indices of these soils also increase in this sequence. The magnitude of dielectric dispersion increases with an increase in the percentage of clay in sand-clay mixtures (Arulanandan, Basu and Scharlin, 1973, Alizadeh, 1975) as shown in Figures 9 and 12 as does the compression index, as is widely known. Olson and Mesri (1970) have shown that the compression index of kaolinite is decreased when the electrolyte concentration is increased from 0.0001 N Sodium to 1.0 N Sodium, and results presented in Figure 12 (Gu and Arulanandan, 1980) show that $\Delta\varepsilon_0$ also decreases with increasing electrolyte concentration.

Since the clay mineral type and amount and pore fluid composition influences the compression index and the magnitude of dielectric dispersion ($\Delta\varepsilon_0$), a correlation between $\Delta\varepsilon_0$ and λ (slope of the consolidation line in e -ln p space) is also expected. The relationship between $\Delta\varepsilon_0$ and λ is shown in Figure 13. This relationship was established mainly for clay minerals.

The influence of particle size on compressibility characteristics has been investigated by Mitchell (1960) and Olson and Mitronovas (1962). Their results show that particle size has an influence on the compressibility of sand-clay mixtures. The influence of particle size on λ was investigated using three soils consisting of 3.3%, 5%, and 10% montmorillonite mixed with 96.7%, 95% and 90% Monterey 10/30' sand of particle size

0.3 mm. The measured values of λ and $\Delta\varepsilon_0$ for these soils are also plotted in Figure 13. It can be seen that the relationship obtained for montmorillonite-sand mixtures is lower than that obtained for clay minerals. This difference is considered to be due to the mechanical interaction between the particles. It can be seen that the values of λ increase with an increase in the amount of clay mineral. This is possibly due to the reduction in the mechanical interaction between particles. The values of λ and $\Delta\varepsilon_0$ were also measured for silt size silica flour (particle size = 0.01 mm) mixed with 3.3% and 5% montmorillonite respectively. The results presented in Figure 13 show that the measured values of λ and $\Delta\varepsilon_0$ for this mixture follow the relationship established for clay minerals. Thus the relationship established for clay minerals is valid for mixtures of clay mineral and non-clay minerals having particle sizes less than 0.01 mm. For soils containing particle sizes greater than 0.01 mm different relationships similar to that developed for Monterey sand and montmorillonite mixtures as shown in Figure 13 have to be established. The slope of the unload-reload line, κ, can also be determined using the concept of inter-cluster and intra-cluster void ratio as described in Arulanandan, 1987.

THE APPLICABILITY OF THE METHOD FOR THE EVALUATION OF SWELL POTENTIAL OF SOILS

Terzaghi (1931) and later Komornik and David (1969) have presented convincing evidence to show that both mechanical and physico-chemical factors control the swelling potential of a soil. Seed et al., (1962) have shown that the swelling potential of soil is mainly controlled by the type and amount of clay mineral. The magnitudes of dielectric dispersion is also controlled by the above factors (Arulanandan and Smith, 1973). Compositional properties and the percentage of swell of several soils are given in Table 2 (Basu and Arulanandan, 1973). The magnitude of dielectric dispersion determined on these soils at saturation are also given in Table 1. Each soil given in Table 1 was consolidated from a slurry under a pressure of 1 kg/cm^2, and a cylindrical specimen approximately 1 inch high and 1.4 inch in diameter was allowed to dry at 65°F and at 50 percent relative humidity. Each dried sample was then confined laterally and allowed to swell freely with no surcharge. Water was introduced by placing the soil on a porous stone that remained saturated with water during the test. Percentage swell given in Table 1 is the total swell expressed as a percentage of the initial height of the sample.

A correlation between the total percentage of swell and the magnitude of dielectric dispersion, $\Delta\varepsilon_0$, measured on samples at the end of the swell test is presented in Figure 14. It may be seen that a good correlation between the percentage of swell and $\Delta\varepsilon_0$ for artificial and natural soils exists. This type of relationship can be used to evaluate the swell potential of soils once the magnitude of dielectric dispersion of saturated samples are known.

SUMMARY AND CONCLUSIONS

A theoretical basis to describe electrical dispersion including the concept of surface conductance has been provided. The electrical response characteristics of saturated soils have been studied. It has been shown that the dielectric constants, ε, and conductivity, σ, in the radio-frequency range (10^6 to 10^8 Hz) vary with the frequency for clay-water-electrolyte systems. These values, ε and σ, are independent of frequency for sand-water-electrolyte systems. The manner in which ε and σ vary as a function of frequency (dielectric dispersion) is dependent on a number of factors. The principle factor influencing the dielectric dispersion behavior is the mineral-solution interface (double layer) characteristics as influenced by the mineralogy and the pore fluid composition. The influence of mineralogy, pore fluid composition and clay content on dielectric dispersion have been shown. In almost all instances, reasonable interpretation of the results in terms of conduction through heterogeneous media is provided by incorporating the surface conductance and particle orientation effects into the electromagnetic theory developed by Maxwell (1881), for non-conducting spherical particles, and Fricke's extension of the electromagnetic theory (1924, 1953) for non-conducting particles of different shapes oriented in a given direction.

Present methods for the determination of soil composition are destructive in that samples have to be remolded before the test. The determination of AC electrical response characteristics is non-destructive and fast. The results of such measurements are capable of quantifying the composition of soils and provide a rapid non-destructive method of characterizing soils. It has been shown that the magnitude of dielectric dispersion, $\Delta\varepsilon_0$, is a compositional index which may be used to evaluate the swell potential and compression index. It is suggested therefore that this technique may provide an additional useful tool for the non-destructive characterization of soil composition and evaluation of the engineering properties of soils.

ACKNOWLEDGMENTS

The support provided by the University of California, Davis Faculty Research Grants, the National Science Foundation Grants and invaluable help provided by Li Xiang in the development of the laboratory electronic system for the dielectric measurements are acknowledged. The authors wish to thank Professor J.K. Mitchell for reviewing the manuscript and making valuable suggestions as well as M.A. Fagan, J.A. Hiscox, and G. P. Broughton for assisting in the preparation of the final manuscript.

Appendix I: REFERENCES

Alizadeh, A. (1975). "Amount and Type of Clay and Pore Fluid Influences on the Critical Shear Stress and Swelling of Cohesive Soils." Thesis submitted in partial satisfaction of the requirements for the degree of Doctor of Philosophy in Engineering, University of California, Davis.

Arulanandan, K. (1966). "Electrical Response Characteristics of Clays and Their Relationships to structure," thesis submitted in partial satisfaction of the requirements for the degree of Doctor of Philosophy in Engineering, University of California, Berkeley.

Arulanandan, K. (1991). "Dielectric Method for the Prediction of Porosity of Saturated Soils." Journal of Geotechnical Engineering, Vol. 117, No. 2, February 1991.

Arulanandan, K. (1987). "Non-destructive Characterization of Particulate Systems for Soil Classification and In Situ Prediction of Soil Properties and Soil Performance." Keynote Lecture, Proceedings of the International Conference in Geotechnical Engineering, Calgary, Canada; A.A. Balkema, Rotterdam, 1987.

Arulanandan, K. (1974). "Properties and Behavior of Marysville Soil" - unpublished results, UCD report

Arulanandan, K., (1973). "Dielectric Properties of Dry Soil Particles" - unpublished results, Report: Dept. of Civil and Environmental Engineering, University of California, Davis.

Arulanandan, K. and Arulanandan, S. (1985). "Dielectric Methods and Apparatus for In Situ Prediction of Porosity and Specific Surface Area (i.e. Soil Type) and for the Detection of Hydrocarbons, Hazardous Waste Materials, and the Degree of Melting of Ice and to Predict the In Situ Stress-Strain Behavior." U.S. Patent.

Arulanandan, K., Basu, R. and Scharlin, R.J. (1973). "Significance of the Magnitude of Dielectric Dispersion in Soil Technology." Highway Research Record, No. 426, pp. 23-32.

Arulanandan, K. and Smith, S.S. (1973). "Electrical Dispersion in Relation to Soil Structure." Journal of the Soil Mech. and Foundation Divs., ASCE, Vol. 99, No. SM12, Proc. Paper 10235.

Basu, R., and Arulanandan, K. (1973). "A New Approach for the Identification of Swell Potential of Soils." Proceedings of the Third International Conference on Expansive Soils, Haifa, Israel, Vol. 1, pp. 1-11.

Casagrande, A. (1948). "Classification and Identification of Soils." Trans. ASCE, 113, pp. 901-992.

Cremers, A., Van Loon, J., and Laudelout, H. (1966). "Geometry Effects for Specific Electrical Conductance in Clays and Soil." Fourteenth National Conference on Clays and Clay Minerals, pp. 149-162.

Dafalias, Y.F. and Arulanandan, K. (1979). "Electrical Characterization of Transversely Isotropic Sands." Archives of Mechanics, 31, 5, pp. 723-739, Warsaw.

Dafalias, Y.F. and Arulanandan, K. (1983), "The Formation Factor Tensor in Relation to Structural Characteristics of Anisotropic Granular Soils," Colloques Internationaux du CNRS, pp. 183-197.

Fricke, H. (1924). "A Mathematical Treatment of the Electrical Conductivity and Capacity of Dielectric Dispersive Systems." Physical Review, Vol. 24, pp. 575-587.

Fricke, H. (1953). "The Maxwell-Wagner Dispersion in a Suspension of Ellipsoids." Journal of Physical Chemistry, Vol. 57, pp. 934-937.

Fricke, H. and Curtis, H.J. (1937). "The Dielectric Properties of Water-Dielectric Interphases." Journal of Physical Chemistry, Vol. 41, pp. 729-745.

Gu, Y. Z. and Arulanandan, K. unpublished UCD report on the Influence of Salt Concentration on the Dielectric Dispersion Behavior of Sand-Clay Mixtures, 1980.

Holtz, W. G. and Gibbs, H. J. (1956) "Engineering Properties of Expansive Clays." Transactions, ASCE, Vol. 121, pp. 641-677.

Komornik, A. and David, D. (1969). "Prediction of Swelling Pressure of Clays." Journal of the Soil Mech. and Foundation Division, ASCE, Vol. 95, No. SM1, pp. 209-225.

Lambe, T.W. and Martin, R.T. (1953-1957). "Comparison and Engineering Properties of Soil." Highway Research Board Proceedings, I-1953, II-1954, III-1955, IV-1956, and V-1957.

Lambe, T.W. and Whitman, R.V. (1969). Soil Mechanics. Wiley, New York.

Maxwell, J.C. (1881). A Treatise on Electricity and Magnetism, 2nd Edition, Clarendon Press, Oxford, p. 398.

Meegoda, N.J. (1985). "Fundamental Characterization of Soils for the Development of an Expression for Permeability for Application in In Situ Testing," thesis submitted in partial satisfaction of the requirements for the degree of Doctor of Philosophy in Engineering, University of California, Davis; submitted for publication.

Mitchell, J.K. (1960). "The Application of Colloidal Theory to the Compressibility of Clays." Proc., Seminar on Interparticle Forces in Clay-Water-Electrolyte Systems, Commonwealth Scientific and Industrial Research Organization, pp. 292-297, Melbourne, Australia.

O'konski, C. T. (1960). "Electrical Properties of Macromolecules, V. Theory of Ionic Polarization in Polyelectrolytes," Journal of Physical Chemistry, Vol. 64, pp. 605-619.

Olson, R.E. and Mesri, G. (1970). "Mechanisms Controlling Compressibility of Clays." ASCE, J. of Soil Mech. and Found. Div., SM6, pp. 1863-1878.

Olson, R.E. and Mitronovas, F. (1962). "Shear Strength and Consolidation Characteristics of Calcium and Magnesium Illite." Proc. 9th Nat. Conf. Clays and Clay Minerals, pp. 185-209.

Sargunam, A. (1973), "Influence of Mineralogy, Pore Fluid Composition and Structure on the Erosion of Cohesive Soils," Dissertation submitted in partial satisfaction of the requirements for the Degree of Doctor of Philosophy in Engineering. University of California, Davis.

Schwan, H.P., Schwarz, G., Maczuk, J. and Pauly, H. (1962). "On the Low Frequency Dielectric Dispersion of Colloidal Particles in Electrolyte Solutions." Vol. 60, pp. 2626-2642.

Seed, H.B., Woodward, R.J. and Lundgren, R. (1962). "Prediction of Swelling Potential for Compacted Clays." Journal of Soil Mechanics and Foundation Division, ASCE, Vol. 88, No. SM3, pp. 53-87.

Skempton, A.W. (1953). "The Colloidal Activity of Clay". Proceedings of the Third International Conference on Soil Mechanics and Foundation Engineering, Vol. I, pp. 57-61.

Terzaghi, K. (1931). "The influence of Elasticity and Permeability on the Swelling of Two Phase Systems." In Colloid Chemistry, Vol. III, J. Alexander, ed., Chemical Catalog Co., Inc., NY, pp. 63-88.

U.S. Bureau of Reclamation (1963). Earth Manual. 1st Edition, revised, Washington, D.C., 783 pp.

Wagner, K. W. (1914). Archives of Electrotechnology, Vol. 2, p. 371.

Weiler, R.A. and Chaussidan, J. (1968). "Surface Conductivity and Dielectrical Properties of montmorillonite Gels." Clays and Clay Minerals, Vol. 16, pp. 147-155, Pergamon Press.

Yogachandran, C. (1988). "Dielectric Method for Laboratory and In Situ Soil Characterization." Thesis submitted in partial satisfaction of the requirements for the Degree of Master of Science in Engineering, University of California, Davis.

Appendix II: Notation

$\Delta\varepsilon_0$ = magnitude of dielectric dispersion

ε_∞ = dielectric constant at high frequency when the dielectric dispersion levels off.

a,b,c = semi axes of the ellipsoids

A_a, A_b = shape indices

f = frequency

f_v, f_h = vertical and horizontal shape factors

F_v, F_h = vertical and horizontal formation factors

k = complex conductivity

k_1, k_2 = complex conductivities of the pore fluid and particle

k_v, k_h = complex conductivities of the mixture in the vertical and horizontal direction

n = porosity

$p(\theta)$ = probability density function

P_θ = orientation factor

R = axial ratio

ε	=	dielectric constant of soil-water mixture
$\varepsilon_1, \varepsilon_2$	=	dielectric constants of pore fluid and particle
$\varepsilon_v, \varepsilon_h$	=	dielectric constants of mixture in the vertical and horizontal direction
σ	=	conductivity of soil-water mixture
σ_1, σ_2	=	conductivities of pore fluid and particle
σ_v, σ_h	=	conductivities of mixture in the vertical and horizontal direction
θ	=	orientation of semi-axis 'a' w.r.t. the vertical axis
ω	=	angular frequency
ρ_s	=	density of particle
σ_s	=	surface conductivity of the particle
λ	=	slope of the consolidation line in the e-lnp space
λ_s	=	surface conductance
SSA	=	specific surface area, area/mass
τ_c	=	critical shear stress

Appendix III: List of Figures

Table 1. Composition and Swell Properties of Natural and Artificial Soils (After Basu and Arulanandan, 1973)

Soil Type	CLAY MINEROLOGY (%)					PI (%)	%<2μm	Activity (=PI/C)	$\Delta\varepsilon_0$	%Swell
	Montmor-illonite	Mixed Layer	Illite	Kaolinite	Non-Clay Minerals (%)					
1*	0	5	0	0	95	6	23	0.26	10	3
1PB*	0	52	5	0	43	31	44	0.70	46	24
2*	0	10	0	0	90	NP	8	0.00	9	2
3*	10	5	0	0	85	1	8	0.13	10	2
3SC*	35	5	0	0	60	30	34	0.88	42	27
6*	0	13	0	0	87	23	25	0.92	27	9
7*	0	5	0	0	95	4.5	10	0.45	18	2
8*	0	20	0	0	80	2	10	0.20	22	2
9*	20	5	0	0	75	22	50	0.44	31	11
11*	0	20	0	0	80	9	19	0.45	11	-
12*	0	10	0	0	90	NP	8	0.00	4	-
15*	15	0	0	0	85	11	33	0.33	23	6
16*	Predominantly Vermiculite					122	48	2.50	16	12
Illite-Grundite	Predominantly Illite					45	50	0.90	-	23
Hydrite R	Predominantly Kaolinite					30	100	0.30	18	4
Marrysville Red (MR)*	Montmorillonite-Kaolinite-Chlorite Mixture					11	20	0.55	41	30

Notes:

* Natural Soils
1. The percentage of different clay minerals present in each soil (shown above) are based on the total weight of the soil. (Clay minerology data supplied by California Division of Highways).
2. Data of Marysville Red is from Arulanandan's personal file.
3. NP = Non plastic

Table 2. Observed and Optimized Electrical Parameters for Given Conditions of Particle Shape, Orientation, Porosity, and Specific Surface Area Used for Calculation of Surface Conductance

Soil Type	Conductivity of Pore Fluid σ_1 (mho/cm) (optimized)	Conductivity of particle σ_2 (mho/cm) (optimized)	Axial Ratio R (assumed)	Orientation Parameter P_θ (optimized)	Porosity n (measured)	Magnitude of Dielectric Dispersion $\Delta\varepsilon_0$	Specific Surface Area SSA (m²/g) (calculated)	Surface Conductance λs (mho) (Calculated)
Figure 9 - Influence of Clay Content on Surface Conductance and Dielectric Dispersion								
40% Montmorillonite + Sand	0.00050	0.01200	40	0.19	0.746	110	241	2.96923E-09
30% Montmorillonite + Sand	0.00039	0.00700	30	0.18	0.709	89	181	2.32496E-09
20% Montmorillonite + Sand	0.00041	0.00520	20	0.17	0.629	55	121	2.62521E-09
Figure 11a and 11b - Influence of Mineralogy on Dielectric Dispersion								
40% Montmorillonite + Sand	0.00050	0.01200	40	0.19	0.746	110	241	2.96923E-09
66% Illite + Sand	0.00040	0.00280	20	0.35	0.480	24	99	1.7277E-09
Kaolinite	0.00020	0.00078	10	0.40	0.623	20	20	2.49582E-09
Silica Flour	0.01150	0	1	0.00	0.530	0	1	0

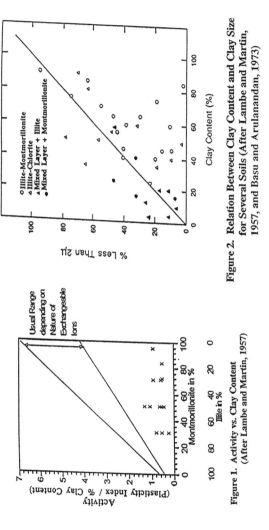

Figure 2. Relation Between Clay Content and Clay Size for Several Soils (After Lambe and Martin, 1957, and Basu and Arulanandan, 1973)

Figure 1. Activity vs. Clay Content (After Lambe and Martin, 1957)

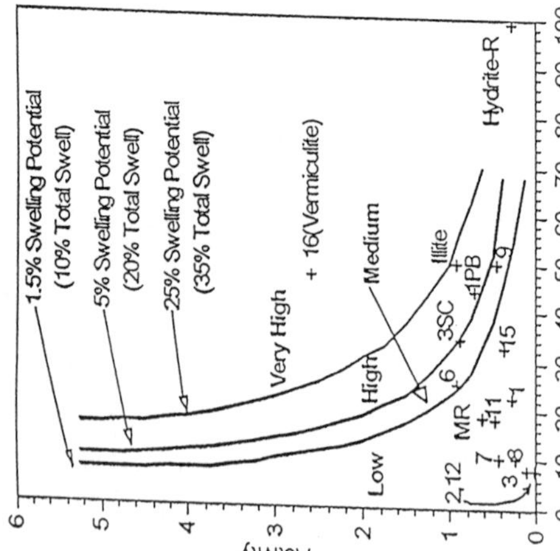

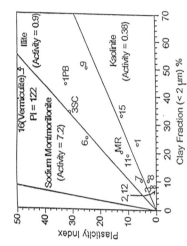

Figure 4. Clay Fraction and Plasticity Index of Some Natural Soils in Relation to the Activity Chart of Skempton (1953).

Figure 3.   Classification Chart for Swelling Potential (after Seed et al., 1962)

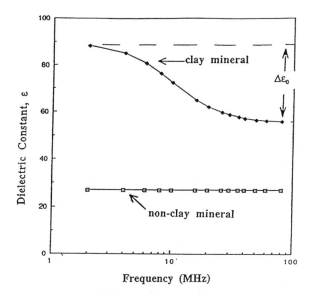

Figure 5.   Typical Dielectric Dispersion Behavior
of a Clay Mineral and a Non-clay Mineral

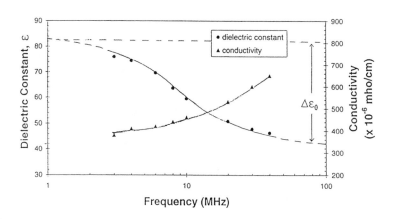

Figure 6.   Electrical Dispersion Characteristics of Marysville Soil

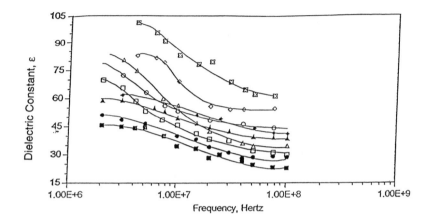

**Figure 7. Dielectric Dispersion Behavior of Several
Natural Soils (Sargunam, 1973)**

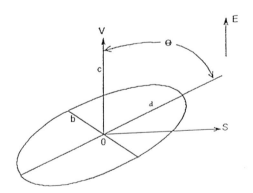

Figure 8. Schematic Illustration of a Particle's
Orientation in Transversely Isotropic
Soils

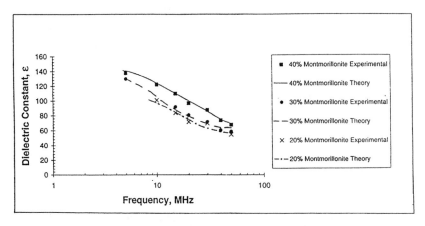

**Figure 9. Comparison of Theoretical and Experimental
Dielectric Dispersions of Montmorillonite Soils
Mixed With Different Percentages of Sand**

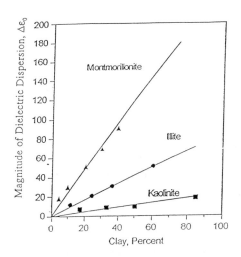

Figure 10. Effect of Clay Type and Amount
on the Magnitude of Dielectric
Dispersion (Alizadeh, 1975)

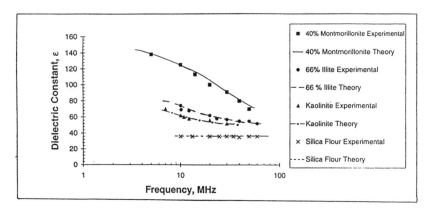

Figure 11a. Comparison of Theoretical and Experimental Variation of Dielectric Constant of Different Clay Minerals as a Function of Frequency.

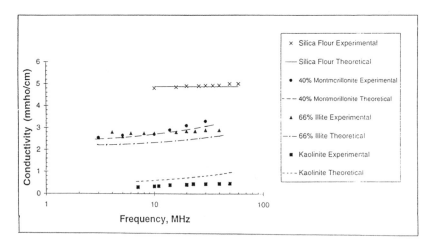

Figure 11b. Comparison of Theoretical and Experimental Conductivities of Different Clay Minerals as a Function of Frequency

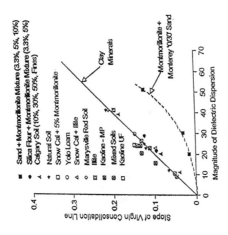

Figure 13. Correlation between the Slope of the Isotropic Consolidation Line e-lnp' Space (λ) and the Magnitude of Dielectric Dispersion ($\Delta\varepsilon_0$).

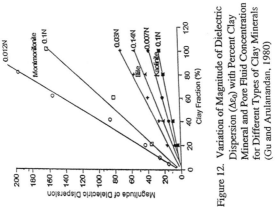

Figure 12. Variation of Magnitude of Dielectric Dispersion ($\Delta\varepsilon_0$) with Percent Clay Mineral and Pore Fluid Concentration for Different Types of Clay Minerals (Gu and Arulanandan, 1980)

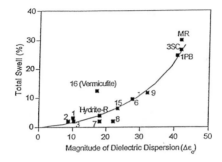

Figure 14. Relationship between Percent Total Swell
and the Magnitude of Dielectric Dispersion

RADIO FREQUENCY DIELECTRIC DISPERSION
WITH ENGINEERING APPLICATIONS

By Scott S. Smith, [1]M. ASCE and K. Arulanandan, [2]M. ASCE

INTRODUCTION

Over the past several decades, powerful analytical and numerical tools have become available for analyzing geotechnical problems. However, the ability to characterize the in-situ physical properties of soils as input to these powerful tools has not significantly advanced. In many cases blow counts associated with driving a sampler into a soil are still used as an indirect measure of a soil engineering parameter as was done decades ago. Clearly future advances in the ability of geotechnical engineers to perform more accurate analyses will be dependent on the development of methods to more accurately characterize in-situ properties. The most beneficial characterization method would be performed in a bore-hole so as to eliminate the need for obtaining the always difficult, if not impossible, "undisturbed" sample that must be tested in a laboratory by some destructive test method after its in-situ stresses have been altered.

As an initial step in the assessment of possible technologies that could eventually be utilized in the development of down-hole in-situ characterization using electrical techniques, the authors investigated the applicability of an electrical model to characterize a soil. The initial research presented herein concentrated on saturated, fine-grained soils. The methodology is based on the variation of the apparent dielectric constant, ϵ', and conductivity, σ, with change in alternating current frequency in the radio frequency range (10^6 Hz $- 10^8$ Hz). This phenomena is termed dielectic dispersion and has been shown to characterize the mineralogy, state of packing, and area of particle contact in fine grained soils (3,4, 6, 16).

[1]Vice President, Harding Lawson Associates, Reno, Nevada

[2] Prof. of Civ. Engrg., Univ. of California, Davis, Calif. 95616.

Fig 1 illustrates a dispersion curve for saturated illite clay that was obtained in the laboratory.   As can be seen, ϵ' decreases and σ increases with increasing frequency.   Previous work has demonstrated that the principal factors controlling the dielectric dispersion of saturated fine grained soil in the radio frequency range are the composition properties of the liquid and solid phase (e.g., mineralogy) and the heterogeneous nature (e.g., state of particle packing, area of particle contact) (3, 16).   Engineering properties of fine grained soils are controlled by these same factors.

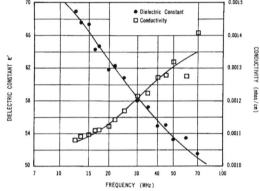

FIG. 1.—Dielectric Constant and Conductivity Dispersion Curves of Illite Clay

In an effort to develop a nondestructive technique for predicting engineering properties of soil in-situ the writers utilized a simplified three element heterogeneous electrical model (Fig. 2) to simulate a complex soil system (15).

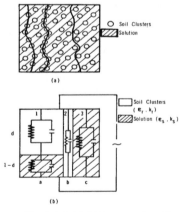

FIG. 2.—Heterogeneous Electrical Model: (a) Representation of Current Path through Soil; (b) Three Element AC Model

In Fig. 2, each zone is represented by a parallel circuit of resistor and capacitor. Impedance of each zone is determined by its dimensions (a, b, c, d) and the specific conductivity (k $_r$, k$_s$) and dielectric constant (ϵ_r,, ϵ_s) of the material forming the zone. The parameters of the model are both compositional and heterogeneous, and the model exhibits a theoretical apparent dielectric and conductivity dispersion in the radio frequency range.

The schematic diagram shown in Fig. 2 (a) assumes the flow of electrical current through the saturated soil can have three paths: (1) Through solution and clusters of particles in series; (2) through clusters in contact with each other; (3) through solution only. Fig. 2 (b) is a schematic representation of the electrical model providing the same three paths. The geometrical parameters a, b, and c represent the fractional cross section of the three paths and d represents the length of electrical current flow through clusters in path (1). The impedance of this model is uniquely determined by the geometrical parameters and the dielectric constants (ϵ_r, ϵ_s) and conductivities (k$_r$,k$_s$) of the clusters and solution, respectively. In the case of soils saturated with water ϵ_s is approximately 79 and k$_s$ is approximately equal to the conductivity of the pore fluid. The apparent dielectric constant, ϵ_{th} and the apparent conductivity, σ_{th}, of the model have been evaluated as a function of frequency by elementary electrical network analysis (13) and are provided in equation 1, 2 and 3:

$$\epsilon'_{th} = \frac{a}{d(1-d)S}\left[\frac{\epsilon_r k_s^2}{1-d} + \frac{\epsilon_s k_r^2}{d} + \omega^2 \epsilon_v^2\left(\frac{\epsilon_r \epsilon_s^2}{1-d} + \frac{\epsilon_r^2 \epsilon_s}{d}\right)\right] + b\epsilon_r + c\epsilon_s, \quad (1)$$

$$\sigma_{th} = \frac{a}{d(1-d)S}\left[\frac{k_s k_r^2}{1-d} + \frac{k_r^2 k_s}{d} + \omega^2 \epsilon_v^2\left(\frac{\epsilon_s^2 k_r}{1-d} + \frac{\epsilon_r^2 k_s}{d}\right)\right] + bk_r + ck_s, \quad (2)$$

$$\text{in which } S \equiv \left(\frac{k_s}{1-d} + \frac{k_r}{d}\right)^2 + \omega^2 \epsilon_v^2\left(\frac{\epsilon_d}{1-d} + \frac{\epsilon_r}{d}\right)^2 \cdots\cdots\cdots (3)$$

and $\epsilon_v =$
the capacitance of the unit capacitor in vacuum, i.e., 0.0885 x 10 $^{-12}$ farads; k$_r$ and k$_s$ are given in mho/cm; ω = the angular frequency in sec^{-1}.

These equations can be fitted to experimental frequency dispersions of the apparent dielectric constant and conductivity of saturated fine grained soils by computer optimization of the geometrical (a, b, c, d) and compositional (ϵ_r, ϵ_s, k$_r$, k$_s$) parameters. By systematically varying the structure determining factors within the fine grained soils, the ability of the parameters of the model to characterize the composition and heterogeneous nature of the soil system has previously been demonstrated (3, 16). The c parameter was found to be a measure of the size and distribution (tortuosity) of the intercluster pores and the intercluster void ratio. The parameter is very sensitive to intercluster void ratio and intercluster pores and independent of compositional variations. Model parameter b is a measure of the contact area between the clusters. In the case of untreated fine grained soils, this parameter is always small, as would be expected. The term ϵ_r is mainly an average measure of the type of clay mineral present in the soil and the intracluster void ratio. The distinctive ϵ_r values obtained for different wet minerals is a result of each mineral's characteristic water retention. The parameter is thus referred to as the dielectric constant of the wet soil cluster.

The observed variation of ϵ_r with water content and fabric (3, 16) substantiates the assumption that the solid phase is composed of aggregates of primary particles. The heterogeneous model is, therefore, in agreement with recent developments and results reported in soil technology. These studies have shown that various particle and particle group arrangements exist in soil. Yong and Sheeran (18) recently presented a scheme of fabric classification based upon the interaction of groupings of particles. These fabric units are termed domains, clusters, and peds. The soil response behavior involves the interaction of fabric units of various sizes. Barden (7) covered the current evidence supporting a similarity in the bulk behavior of clay and sand. This concept is predicated on the belief that groups of particles rather than individual particles interact in clays.

The fundamental nature of the electrical dispersion in fine grained soils and the ability to determine the numerical parameters $b,$ $c,$ and ϵ_r of the heterogeneous model present an effective tool for studying the engineering behavior of soils in a fundamental quantitive manner. This paper presents a summary of the pertinent results and conclusions of three such studies. The studies are considered separately and included: (1) The evaluation of the heterogeneous model's ability to monitor structural changes during the hydration of soil cement; (2) relationship between model parameter ϵ_r and swell potential; and (3) relationship between model parameter c and hydraulic permeability. Other similar studies are discussed for comparison purposes.

EXPERIMENTAL PROCEDURE

The basic equipment and procedures followed in these studies have been reviewed in detail in previous publications (3, 6, 16). The electrical dispersion measurements of the soils were performed with a type 250 RX meter (Booton Radio Corporation, Division of Hewlett-Packard, Rockaway, N.J.). The instrument is essentially a Schering bridge, with oscillator, amplifier detector, and null-indicator designed to measure equivalent parallel conductance and capacitance in the range 0.0 mhos-0.067 mhos at frequencies of 0.5 megacycles-

2.50 megacycles. The soils to be measured were loaded into an electrical cell and connected to the electrical bridge.

A digital computer optimization program (1, 12) was employed to analyze the emperical dispersion curves. The optimization program determines the values of the geometrical and compositional parameters of the electrical model such that the frequency dispersion curves of the apparent dielectric constant and conductivity calculated from the theoretical Eqs. 1 and 2 fit the empirical results within acceptable error limits. The value of ϵ_S was kept constant at 79, the dielectric constant of water, during all computer runs. Previous work has demonstrated the optimized model parameters give reasonable values from our knowledge of soil science and physical science.

STRUCTURAL CHANGES OF SOIL CEMENT DURING HYDRATION

Mitchell and El Jack (11) have utilized the electron microscope to investigate changes in fabric and composition of soil cement during hydration. The heterogeneous model was applied to one of the soil cements studied by Mitchell and El Jack to evaluate the model's ability to monitor these structural changes.

The soil cement was commercially available kaolinite clay (Hydrite UF) with 24% Type II portland cement by weight. The dry mixture of kaolinite and cement was mixed with distilled water to a saturated water content of 140%. A high water content was necessary to assure that no air voids would result when the soil cement was loaded into the electrical measuring cells. Electrical measurements were performed at different time intervals as curing proceeded: 1 day, 2 days, 8 days, 16 days and 32 days. The measured dielectric and conductivity dispersions for these curing periods are given in Figs. 3 and 4, respectively.

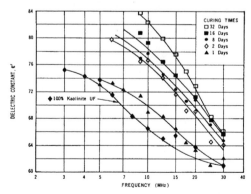

FIG. 3.—Experimental Dielectric Constant Dispersion Curves for Kaolinite Hydrite UF Soil Cement at Varying Times after Mixing and for Pure Kaolinite Hydrite UF

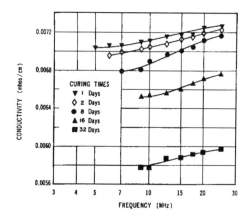

FIG. 4.—Experimental Conductivity Dispersion Curves for Kaolinite Hydrite UF Soil Cement at Varying Curing Times

For an indication of initial soil structure at time zero, electrical dispersion measurements were also performed on pure kaolinite (Hydrite UF) mixed with distilled water to a water content of 140%. It was necessary to use pure Kaolinite instead of the kaolinite-cement mixture because the electrical properties of the soil cement during the first day of curing were found to change too rapidly for the measurement of an electrical dispersion. Although the pure kaolinite is not an exact measure of the soil cements' initial structure, it does give an approximate picture of the particulate structure of the soil cement at initial mixing

The curve fitting computer program was applied to the experimental curves. A typical curve fitting result is presented in Fig. 5 for the case of the 32-day curing period. The optimized parameters values are listed in Table 1.

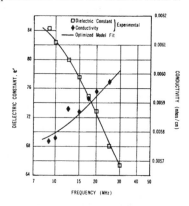

FIG. 5.—Curve Fitting of Theoretical Electrical Model Dispersion with Experimental Dispersion of Kaolinite Hydrite UF Soil Cement for 32-Day Curing Period

TABLE 1.—Optimized Model Parameter Values during Hydration of Soil Cement

Material description (1)	Initial water content, as a percentage (2)	Pore fluid extract conductivity, in mhos per centimeter (3)	Optimized Model Parameters						
			a (4)	b (5)	c (6)	d (7)	$\epsilon_r{}^{a}$ (8)	k_r (9)	k_s (10)
Kaolinite UF	140	0.00006	0.33	0.02	0.65	0.38	9.	2.3×10^{-4}	0.00012
Soil cement— 1-day hydration	140	0.007	0.02	0.33	0.65	0.01	13.	3.1×10^{-6}	0.011
Soil cement— 2-day hydration	140	0.007	0.03	0.34	0.63	0.01	12.	2.9×10^{-6}	0.011
Soil cement— 8-day hydration	140	0.007	0.04	0.36	0.60	0.01	13.	1.6×10^{-5}	0.011
Soil cement— 16-day hydration	140	0.007	0.05	0.37	0.58	0.01	12.	1.6×10^{-5}	0.011
Soil cement— 32-day hydration	140	0.007	0.06	0.39	0.55	0.01	11.	1.5×10^{-5}	0.010

[a] Model parameter ϵ_r was kept constant at 79.0 during optimization.

The c parameter is seen to decrease with increasing hydration time of the soil cement. This reflects the reduction in unrestricted pore paths through the soil cement due to the interlocking of particles by gel that results from the hydration process. This is in agreement with the observed decrease in the coefficient of permeability of clay soils upon the addition and curing of cement (13).

The optimized b parameter of the soil-cement samples is seen to be significantly greater than that of the pure kaolinite; indicating that the cement and clay particles in the soil cement mixture are no longer discrete. The area of contact between solids is quite large indicating the fabric is more of a solid network.

The relationship between the b parameter and the curing time is plotted in Fig. 6. The major increase in b takes place during the first day of curing. After the third day, the rate of increase of b is small but measurable, and is seen to be almost constant. The large increase in b during the first day can be ascribed to the hydration of the cement. Gel resulting from the hydration interlocks the discrete soil particles into a more solid network. The small but constant rate of increase of b after the third day may be due to secondary cementing compounds resulting from the reaction of the clay particles with the lime liberated during hydration.

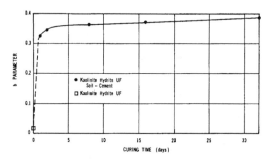

FIG. 6.—Relationship between *b* Parameter of Electrical Model and Curing Time of Kaolinite Hydrite UF Soil Cement

Taylor and Arulanandan (17) have utilized radio frequency electrical dispersion and the same electrical three element model to investigate structural changes during curing of pure cement pastes. Water – cement ratios ranged from approximately 0.3 to 0.4. Electrical measurements were made at different time intervals up to 1 week to observe changes as curing progressed. The optimized *b* parameter, representing solid to solid contact, followed the same pattern as observed above for the soil cement. The *b* parameter increased rapidly during the first day and then showed a slow but steady increase. The *c* parameter also followed the same pattern as with the soil cement discussed above. It decreased steadily with time as a result of gel formation.

Values of *b* determined by Taylor and Arulanandan for the pure pastes were 50% to 100% greater than the *b* parameters reported herein for the soil cement. This would be the expected result since the greater percentage of cement present in the cement paste would be expected to result in a greater percentage of the volume to be occupied by the gel than in the soil cement.

SOIL EXPANSION

A soil's potential to swell when exposed to water is mainly a function of the type and amount of clay minerals present. Each clay mineral has its own inherent capacity for water absorption. The ϵ_r, parameter represents the dielectric constant of a wet soil cluster and is therefore an average measure of the type of clay mineral. Its value should be between 4 (the dielectric constant of a dry silicate mineral) and 79 (the dielectric constant of pore water) depending on the proportions of the two phases in the cluster and, thus, the water retention characteristics of the clay minerals present. The relationship between the dielectric constant of the wet soil cluster, ϵ_r, and the swell potential of a soil has been investigated (5, 8).

Fourteen soils were studied in the most recent investigation (8).   Ten of the samples were supplied by the state of California, Division of Highways and were sampled from throughout the state.   One of the soils was from the Las Vegas, Nev. area and was supplied by the Federal Housing Administration.   The remainder of the soils are commercially available.

Each soil tested was mixed with distilled water and then consolidated from an initial slurry under a 1 kg/cm^2 load. A cylindrical specimen was allowed to dry in a 50% moisture room. The dried soil specimen was then laterally confined and allowed to swell without surcharge. Radio frequency dispersion measurements were performed on each soil in the consolidated state before drying. The curve fitting computer program was then applied to the resulting empirical dispersion curves and the heterogeneous model parameters were calculated. Table 2 summarizes the optimized model parameter values after consolidation for each soil.

TABLE 2.—Optimized Model Parameter Values for Swell Potential Soil Samples (7)

Soil identification (1)	Water content after consolidation, as a percentage (2)	Optimized Model Parameters [a]						
		a (3)	b (4)	c (5)	d (6)	ϵ_r (7)	k_r (8)	k_s (9)
1[b]	39.	0.76	0.02	0.22	0.30	10.	0.0006	0.0008
1PB[b]	106.	0.61	0.01	0.38	0.99	36.	0.0053	0.0007
3[b]	39.	0.68	0.01	0.31	0.72	18.	0.0018	0.0013
3SC[b]	75.	0.87	0.01	0.12	0.69	41.	0.0028	0.0008
6[b]	64.	0.72	0.01	0.27	0.88	18.	0.0036	0.0015
7[b]	32.	0.66	0.01	0.33	0.75	9.	0.0007	0.0005
8[b]	47.	0.55	0.03	0.42	0.94	8.	0.0015	0.0005
9[b]	50.	0.50	0.03	0.47	0.98	23.	0.0060	0.0007
13[b]	48.	0.49	0.01	0.50	0.98	26.	0.0046	0.0006
15[b]	46.3	0.46	0.03	0.51	0.98	13.	0.0041	0.0006
16[c]	114.	0.67	0.02	0.31	0.88	25.	0.0007	0.0003
Illite (flocculated)	78.	0.59	0.02	0.39	0.90	26.	0.0036	0.0011
Illite (dispersed)	65.	0.90	0.01	0.09	0.99	48.	0.0076	0.0005
Kaolinite Hydrite R	65.	0.67	0.01	0.32	0.23	8.	0.00012	0.0002

[a] Model Parameter ϵ_s was kept constant at 79.0 during optimization.
[b] Natural soils from throughout the state of California; supplied by the state of California Division of Highways.
[c] Natural soil from Las Vegas; supplied by Federal Housing Administration.

The optimized values of ϵ_r are plotted versus the percentage swell for each soil in Fig. 7. A linear curve has been drawn through the data points. A good correlation between increasing dielectric constant of the wet soil cluster and increasing percentage swell was obtained for these various soil types. It can be concluded from the results obtained from these soils that ϵ_r is a measure of a soil's water absorption characteristics and may prove useful for the evaluation of swell potential in soils.

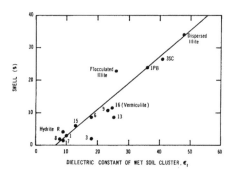

FIG. 7.—Relationship between Swell and Dielectric Constant of Wet Soil Cluster, ϵ_r (7)

Fernando, Smith and Arulanandan (10) have investigated the relationship between the dielectric dispersion characteristics of clay soils and the soils expansion index (EI), as determined using the Uniform Building Code standard. Sixteen natural soils from throughout California were tested. Radio frequency dispersion measurements were performed on each soil at the liquid limit. The authors did not use the three element model. They determined the magnitude of dielectric dispersion, $\Delta\epsilon_0$, defined as the difference in the dielectric constant at about 3×10^6 Hz and 75×10^6 Hz, directly from the dielectric dispersion curve. Figure 8 illustrates the relationship between the measured EI and the magnitude of dielectric dispersion, $\Delta\epsilon_0$. The linear relationship between the two parameters is seen to be quite good. The authors conclude that $\Delta\epsilon_0$ alone is predominately a measure of the amount and type of clay mineral.

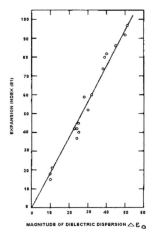

FIG. 8.—Relationship Between Expansion Index and Magnitude of Dielectric Dispersion (10)

Other researchers have used the direct measurement of $\Delta\epsilon_0$ as a measure of the average clay mineral in a soil in order to predict the erodibility of a soil (10). Although not the only parameter needed to predict erodibility, $\Delta\epsilon_0$ was found to be one of a few parameters that effectively characterized a soil's susceptibility to erosion.

HYDRAULIC PERMEABILITY

The hydraulic permeability of cohesionless soils is reasonably well represented by the Kozeny-Carmen equation (9). This equation predicts that for a given saturated soil, the hydraulic permeability is a function of the void ratio exclusively. However, it has been demonstrated by many investigators that the Kozeny-Carmen equation fails to predict measured flow rates in clays. Olsen (14) has shown that the concept of "unequal pore sizes" can explain these discrepancies. He suggested that the primary particles in a clay mass are arranged in clusters and the total porosity of the soil system is distributed among intercluster and intracluster components. The flow channels surrounding the clusters (intercluster pores) will be considerably larger than those within the clusters between the primary particles. By deriving the relationship for flow rates through a cluster model similar to that of Fig. 2 and assuming the fluid only flows through the large intercluster pores around and between the clusters, he was able to explain the unique hydraulic flow characteristics of clays.

The c parameter of the electrical model represents the fraction of the electrical current that passes through the intercluster solution. It has been found to be a measure of the size and distribution of these intercluster pores (4). Since the role of the intercluster pores in controlling permeability has been established, [Olsen (14)], a limited experimental program was conducted to investigate the relationship between the c parameter and the hydraulic permeability of saturated fine grained soils.

Seven clay soils prepared from commercially available clay minerals were tested: five illite (Grundite); one kaolinite(HydriteR); one montmorillonite (Vol-Clay). The illite and kaolinite soils were first made homoionic to a particular cation by mixing a very dilute suspension at 0.1 N and allowing the particle to settle out. This was repeated several times to assure thorough cation replacement by the law of mass action. The final desired pore fluid concentration was obtained by decanting the required portion of the solution left after the particles had settled, adding distilled water, and thoroughly mixing again. The montmorillonite was treated in a similar fashion. However, due to its strong colloidal characteristics, a centrifuge was employed to accelerate the particle settling process. The chloride form of each cation was used. The kaolinite and montmorillonite clay were made homoionic to Na^+. Three of the illite clay soils were made homoionic to Na^+, and the remaining two to K^+ and Li^+ respectively. The illite clay soils were all prepared at an approximate concentration to 0.012 N.

After the desired pore fluid electrolyte concentration had been obtained, the soils were consolidated from a slurry consistency under a load of 1 kg/cm^2 After the soils were through consolidating, they were leached by applying an 8-psi vacuum to the bottom of the sample and introducing an excess of solution to the top. The added solution contained the same electrolyte type and was approximately the same concentration as the pore fluid. By monitoring the volume of water leached through over a period of time, the hydraulic permeabilities of the samples were determined. Each soil was then loaded into the electrical cell and the electrical measurements were performed and the model parameters were determined as previously outlined.

The computed model parameters are presented in Table 3 along with other pertinent soil data. A plot of the c parameter versus the hydraulic permeability is presented in Fig. 9. A best fit line through the data points is shown. Although the number of data points is limited, there appears to be general agreement between increasing c and increasing hydraulic permeability. It can be concluded from these results that the *c* parameter may offer a numerical evaluation of the size and distribution of intercluster pores and could serve as a tool in the measurement and study of the hydraulic permeability of fine grained soils.

TABLE 3.—Optimized Model Parameter Values for Hydraulic Permeability Soil Samples

Material description (1)	Water content, as a percentage (2)	Pore fluid extract conductivity, in mhos per centimeter (3)	Cation type (4)	Hydraulic permeability, in centimeters per second (5)	Optimized Model Parameters*						
					a (6)	b (7)	c (8)	d (9)	ϵ_r (10)	k_r (11)	k_r (12)
Illite	61	0.0014	Na⁺	7.7×10^{-8}	0.66	0.01	0.33	0.66	20.	0.0017	0.0083
Illite	76	0.0009	Na⁺	7.5×10^{-8}	0.65	0.01	0.34	0.58	19.	0.0011	0.0006
Illite	62	0.0014	Na⁺	1.1×10^{-7}	0.64	0.01	0.36	0.65	21.	0.0015	0.0008
Illite	61	0.0011	K⁺	1.1×10^{-7}	0.62	0.02	0.36	0.52	13.	0.0012	0.0008
Illite	60	0.0012	Li⁺	1.3×10^{-7}	0.64	0.01	0.35	0.49	14.	0.0011	0.0008
Kaolinite	60	0.0002	Na⁺	3.0×10^{-7}	0.56	0.01	0.43	0.43	10.	0.0004	0.0002
Montmorillonite	457	0.0010	Na⁺	2.0×10^{-8}	0.82	0.02	0.16	0.93	63.	0.0043	0.0005

*Model Parameter ϵ_r was kept constant at 79.0 during hydration.

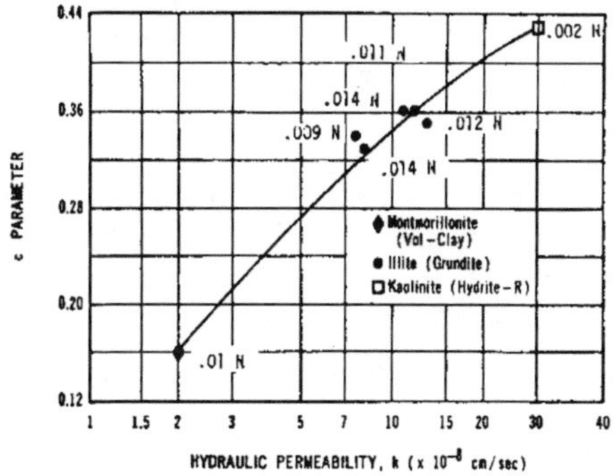

FIG. 9. - Relationship between *c* Parameter of Electrical Model and Hydraulic Permeability

SUMMARY AND CONCLUSIONS

As an initial step in the assessment of possible technologies that could eventually be utilized in the development of down-hole characterization using electrical techniques, the writers have employed a simplified three element electrical model to represent the complicated soil-water system. The model exhibits a theoretical apparent dielectric and conductivity dispersion in the radio frequency range. The parameters of the model are both compositional and heterogeneous (structural). For a particular soil system, numerical values of the model parameters can be determined by computer curve fitting the theoretical equations of the model to the experimentally determined radio frequency apparent dielectric and conductivity dispersion curves of the soil.

In this article three of the model parameters *(b, c,* and ϵ_r) in particular, have been demonstrated to numerically characterize the composition and structure of saturated fine grained soils. The *c* parameter is shown to be a measure of the size and distribution (tortuosity) of the intercluster pores. Parameter *b* is a measure of the contact area between the clusters. The term ϵ_r, was found to represent the dielectric constant of the wet soil cluster and depends mainly on the water retention characteristics of the clay minerals present. It is, therefore, mainly an average measure of the type of clay mineral present.

The fundamental nature of electrical dispersion in fine grained soils and the numerical parameters *b, c,* and ϵ_r, of the model present an effective nondestructive tool for studying the engineering behavior of soils in a fundamental quantitative manner. The pertinent results or discussion of five such studies are presented in this paper. The five studies included the evaluation of: the models ability to monitor structural changes during the hydration of soil cement and cement paste; the relationship between model parameter ϵ_r, and swell potential and between $\Delta\epsilon_0$ and the Expansion Index; and the relationship between model parameter c and hydraulic permeability.

Mitchell and El Jack (9) have utilized the electron microscope to investigate changes in fabric during hydration of soil cement. The heterogeneous model was applied to one of the soil cements studied by Mitchell and El Jack (Kaolinite clay-Hydrite UF). The c model parameter was found to decrease with the hydration of the soil cement. This reflects a decrease in the unrestricted pore paths through the soil cement. The *b* parameter for the soil cement was found to be significantly greater than that of the pure soil. This indicates that the cement and clay particles in the soil-cement mixture are no longer discrete and have formed more of a solid network. The changes in structure of the soil cement, as monitored by the model parameters, were found to agree qualitatively with the changes suggested by Mitchell and El Jack. Results of a similar application of the electrical model with cement paste are also discussed (15a).

The relationship between model parameter ϵ_r, and swell potential has been previously investigated using 11 natural and three artificial soils. A good correlation between percentage swell and ϵ_r, was obtained. It can be concluded from the results obtained from these soils that ϵ_r, is a measure of a soil's water absorption characteristics and may prove useful for the evaluation of swelling potential in soils. Results of an investigation that compared Expansion Index to the magnitude of dielectric dispersion in the radio frequency range is also presented.

A limited experimental program was conducted to investigate the relationship between the c parameter of the model and the hydraulic permeability of saturated fine grained soils. The value of the c parameter was found to increase with the value of the hydraulic permeability. This limited data suggests that the c parameter may offer an effective tool in the measurement and study of the hydraulic permeability of saturated fine grained soils.

ACKNOWLEDGMENTS

The research described herein is part of a continuing investigation into the relationship between electrical and mechanical properties of soils, supported by National Science Foundation Grant Nos. GK-3539 and GK-20372. The support is gratefully acknowledged.

REFFERENCES

1. Arulanandan, K., and Mitra, S. K., "Soil Characterization by Use of Electrical Network," *Proceedings of the 4th Asilomar Conference on Circuits and Systems,*
    Nov., 1970, pp. 480-485.
2   Arulanandan, K., Sargunam, A. Loganathan, P., and Krone, R. B., "Application of Chemical and Electrical Parameters to Prediction of Erodibility", Highway Res. Board Spec. Rep. , No. 135, pp 42-51, 1973

3. Arulanandan, K., and Smith, S. S., "Electrical Dispersion in Relation to Soil Structure," *Journal of the Soil Mechanics and Foundation Division,* ASCE, Vol. 99, No. SM12, Proc. Paper *10235,* Dec., 1973, *pp. 11 13-1133.*
4. Arulanandan, K., and Smith, S. S., "Soil Structure Evaluation by the Use of Radio Frequency Electrical Dispersion," *International Symposium on Soil Structure,* Aug., 1973.
5. Aralanandan, K., Smith, S. S., and Linkhart, T. A., "Characterization of Clays by Electrical Methods," *Proceedings of the Sixth Annual Meeting of Clays and Clay Minerals Society, 1969.*
6. Arulanandan, K., Smith, S. S., and Spiegler, K. S., "Radio Frequency Properties of Polyelectroyte Systems," *Proceedings,* NATO Advanced Study Institute, Forgesles-Eaux, June, 1972.
7. Barden, L., "The Influence of Structure on Deformation and Failure in Clay Soil," *Geotechnique,* Vol. 22, No. 1, Mar., 1972, pp. 159-163.
8. Basu, R., "Identification and Prediction of Swell of Expansive Earth Materials," thesis presented to the University of California, at Davis, Calif., in June, 1972, in partial fulfillment for the degree of Master of Science.
9. Carmen, P. C., *Flow of Gases through Porous Media,* Academic Press, New York, N.Y., 1956.
10. Fernando, Jonas, Smith Robert, and Arulanandan, Kandiah, "New Approach To Determination of Expansion Index", Technical Note, Journ. of the Geotech. Engin. Division, ASCE, Vol. 101, No GT9, pp 1003-1008, September, 1975
11. Mitchell, J. K., and El Jack, S. A., "The Structure of Clay Cement and Its Formation," Institute of   Transportation and Traffic Engineering, University of California, Berkeley, 1965.
12. Muat, R. W., "Driving-Point Characteristic Approximation with R.C. Network," thesis presented to the University of California, Davis, Calif., in 1968, in partial fulfillment of the requirements for the degree of Master of Science.

13. *Nussbaum,* P. J., and Colley, B. E., "Dam Construction and Facing with Soil-Cement," *Research and Development Bulletin,* Portland Cement Association, 1971.
14. Olsen, H. W., "Hydraulic Flow Through Saturated Clays," *Proceedings,* 9th National Conference Clays and Clay Minerals, 1962, pp. 131- 1 6 1.
15. Sachs, S. B., and Spiegier, K. S., "Radiofrequency Measurements of Porous Conductive Plugs, Ion-Exchange Resin-Solution System," *Journal of Physical Chemistry, Vol.* 68, 1964, pp. 1214-1222.
16. Smith, S. S., "Soil Characterization by Radio Frequency Electrical Dispersion," dissertation presented to the University *of California, Davis, Calif.,* in 1971, *in partial* fulfillment of the requirements for the degree of Doctor of Philosophy.
17. Taylor, Michael A., and Arulanandan, K., "Relationship Between Electrical and Physical Properties of Cement Pastes", Cement and Concrete Research, Vol. 4, pp 881-897, 1974
18. Yong, R. N., and Sheeran, D. E., "Fabric Unit Interaction and Soil Behavior," *Proceedings,* International Symposium on Soil Structure, Aug., 1973, pp. 175-183.

NOTATION

The following symbols are used in this paper:

a, b, c, d	=	geometrical parameters of electrical model;
k_r, k_s	=	conductivity parameters of electrical model;
ϵ_r, ϵ_s	=	dielectric constant parameters of electrical model;
ϵ_{th}	=	theoretical apparent dielectric constant from electrical model;
ϵ_v	=	capacitance of unit capacitor *in vacuum;*
ϵ'	=	apparent dielectric constant (experimentally measured);
σ	=	conductivity (experimentally measured);
σ_{th}	=	theoretical conductivity from electrical model; and
ω	=	angular frequency.

Non-destructive Measurement of Soil Liquefaction Density Change by Crosshole Radar Tomography, Treasure Island, California

Robert E. Kayen[1], Walter A. Barnhardt[1], Scott Ashford[2] and Kyle Rollins[3]

ABSTRACT

A ground penetrating radar (GPR) experiment at the Treasure Island Test Site [TILT] was performed to non-destructively image the soil column for changes in density prior to, and following, a liquefaction event. The intervening liquefaction was achieved by controlled blasting. A geotechnical borehole radar technique was used to acquire high-resolution 2-D radar velocity data. This method of non-destructive site characterization uses radar trans-illumination surveys through the soil column and tomographic data manipulation techniques to construct radar velocity tomograms, from which averaged void ratios can be derived at 0.25 - 0.5m pixel footprints. Tomograms of void ratio were constructed through the relation between soil porosity and dielectric constant. Both pre- and post-blast tomograms were collected and indicate that liquefaction related densification occurred at the site. Volumetric strains estimated from the tomograms correlate well with the observed settlement at the site. The 2-D imagery of void ratio can serve as high-resolution data layers for numerical site response analysis.

INTRODUCTION

This paper describes a new non-destructive technique that uses ground-penetrating radar (GPR) to determine changes in soil density due to the effects of liquefaction events. The density state of soil is principally controlled by the depositional history, sediment texture, post-depositional load history, the influence groundwater, and diagenetic changes to the soil fabric (Mitchell, 1976). Even subtle

---

[1] USGS, 345 Middlefield Road, Menlo Park, CA 94025
[2] UCSD, La Jolla, CA
[3] Brigham Young University, Provo, Utah.

variations of environmental state and intrinsic-physical properties can significantly alter the soil density. Estimation of the *in situ* density of sand is typically made indirectly through empirical correlations from standard penetration testing (SPT) and conventional cone penetration testing (CPT). If a large budget is available, the *in situ* density can be determined from laboratory analysis of frozen samples, or from neutron or gamma ray density logging. Soil density-state and texture have a first order influence on the liquefaction susceptibility of soils, and other factors (e.g. particle orientation) are known to have a secondary influence. The application of field cross-hole GPR methods to characterise a potentially liquefiable soil mass is new, although early work on the relation between soil porosity and FM-frequency velocity dates back to laboratory work using time-domain reflectometry of Topp et al. [1980].

Treasure Island is a hydraulically-filled man-made structure in central San Francisco Bay, north of Yerba Buena Island, constructed for the 1939 Golden Gate International Exposition (Figure 1). The island was built in 1936 and 1937 by hydraulically pumping estuarine soil behind a perimeter rip-rap dike (Figure 1). Treasure Island is a National Geotechnical Experimentation Site [NGES], and a considerable data set of geotechnical information is available for the site. The soil profile at the test site consists of hydraulically-placed fill to a depth of 7 m, underlain by Holocene San Francisco Bay Mud and older Pleistocene estuarine and terrestrial deposits. The hydraulic fill is primarily loose silty fine sand and sandy silt (Bennett, 1994). The water table varied from 0 m (standing water) to 1.5 m below the ground surface while we occupied the site.

The GPR-based technique for measuring soil density changes complements a series of 10 full-scale pile-group and cast in steel shell-pile [CISS] load tests performed between December, 1998 and February, 1999. These load tests were led by researchers at the University of California, San Diego and Brigham Young University, and supported by the California Department of Transportation (Caltrans) and NSF. Other field data obtained from the experiment site included soil boring logs, conventional cone-penetrometer (CPT) logs, and shear wave velocity profiles. All testing at Treasure Island was conducted in cooperation with the United States Navy and the City of San Francisco (Figure 1).

The object of this study was to observe the effects of liquefaction on soil density without the overprinting effect of mechanical lateral loading of the pile groups. To achieve this, an array of three PVC-cased boreholes were placed approximately 8 m away from the 4-pile group. The blast charges were then placed between the piles and the borehole array, roughly equidistant from each. As such, the radar site would be subjected to similar blast intensity as the pile-group, but was isolated from pile group loading following the blast. A drill rig augured the boreholes each to a depth of 9 m and laid out the three holes in a 3-4-5 m triangular pattern. The holes were then cased with PVC liner.

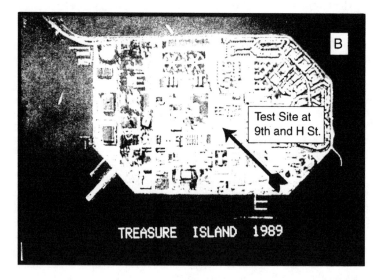

Figure 1. [A] View from north during the hydraulic filling of
Treasure Island, 1937. [B] Treasure Island Naval station at the
time of the Loma Prieta event, 1989.

BLASTING

Blasts consisted of multiple sets of eight 0.5 kg explosive charges placed in circular patterns around the 4-pile group. Each charge pattern had a diameter of approximately 5 meters from the closest pile location, and was approximately 4 meters inside of the radar bore holes (Figure 2). Paired charges were placed approximately 3.5 meters below the excavated surface (approximately 3 meters below the ground water surface) and set off sequentially, with a short blast delay (between 250 and 1000 milliseconds) between charges. This delay was used to maximize the effect of the total blast on the soil mass. For each blast, pore-pressure time-histories were collected in horizontal and vertical pressure transducer arrays. Peak particle velocity and settlements were also measured for the blast. At the 4-pile group, peizometer sensor data indicate that a four meter thick zone liquefied within the soil column. Settlements due to liquefaction measured along three transects across the 4-pile site were at a maximum 16.8 cm., 17.8cm, and 20.7 cm (6.6, 7.0, and 8.2 inches).

RADAR METHODS

Cross hole GPR is a trans-illumination method of survey in which two antennas are lowered down adjacent, parallel boreholes. The transmitter antenna emits a short pulse, or shot, of high frequency (100 MHz in this case) electromagnetic energy. The receiving antenna, located in the other , precisely measures the time required for the signal to travel through the ground, along the plane separating the two boreholes. Travel-times are one-way and measurements are made with picosecond-level (10^{-12} s) precision. Trans illumination involves passing a waveform through the soil to determine the travel time and attenuation characteristics of the wave (Figure 2). As such, cross hole GPR requires careful calibration of the outgoing waveform and shot-time zero to establish the travel time of the signal. In the field, with antennas in fixed positions, multiple wave trace records were recorded and stacked. Stacking involves adding together the waves of the multiple shots (32 shots in our survey) at each point along a profile. Stacking improves data quality by reinforcing the wave signal and suppressing noise. Field wave-trace records are immediately available to the researcher to assess the quality of the survey and individual waveforms.

The accuracy of the travel time measurement is critical for determining the radar velocity. To determine the electronic signal delay, inherent in all circuits, the antennas are held at each borehole and a wave is transmitted through the air. The speed of light in air and the borehole separation are know values, so the required travel time for the wave can be computed. The electronic signal delay is measured as the recorded travel time in excess of that required to cross the borehole separation.

Two different types of cross hole surveys were conducted, a constant offset profile (COP), and a multiple offset gather (MOG). The COP was used to make a quick reconnaissance survey of the site in which both GPR antennas were lowered to

100MHz-Crosshole GPR Survey,
Treasure Island, CA
February 1999

9-PileGroup

transmitter

receiver

Figure 2. Borehole radar investigation at the 9-pile group, Treasure Island. The blast charges were placed between the pile group and radar boreholes.

equal depths down their respective boreholes for each shot (Figure 3a). The COP allows for the rapid collection of radio waves along a horizontal path direction (assumed bedding direction) and equal path length between transmitter and receiver all the way down the soil profile. We used the COP to rapidly identify, in the field, anomalies in travel time and signal strength that would indicate variations in sediment properties. The COP data was also used to distinguish the hydraulic fill from Bay Mud and design a plan for more detailed radar surveys in the fill.

The MOG is a more detailed cross hole survey in which the transmitter antenna is fixed at a certain depth in one borehole while the receiver is moved in regular steps down the other (Fig. 3b). After the receiver collects shots from top to bottom (one complete MOG), the transmitter is lowered a predetermined step-interval and fixed at that new position, and the receiver is again moved down the other hole. The process is repeated until the transmitter reaches the bottom. In this study, we collected a suite of MOGs, each with a step-interval of 0.25 m. Unlike a COP, the path lengths in a MOG vary greatly from shot to shot. For each transmitter-receiver orientation, the path length is computed. In a perfectly homogeneous medium, the first arrivals would form a hyperbola. Deviations from a smooth hyperbolic pattern are indicative of variations in soil properties.

TOMOGRAPHIC MANIPULATION

Substantial computation, both in the field and in post-acquisition processing is required before interpretable images are produced. After wave traces are gathered, it is necessary to pick the first and, if possible, second breaks on each trace. Refracted air waves are common, especially when the antennas are near the surface, and must be distinguished from direct arrivals. This time-intensive process of event picking enables the extraction of travel time, amplitude, and period of the transmitted pulse.

Tomographic analysis utilizes the path length and precise measurements of one-way travel time to determine the velocity structure of the intervening materials. The positions of the transmitter and receiver antennas in the boreholes are well known, and therefore the raypath distance between the two is accurately calculated for each shot. The objective of conducting multiple surveys (COP and MOG) between the same boreholes is to collect travel-time data along as many ray paths and along as many *different angles* as possible. The analysis first divides the single plane connecting the boreholes into a grid of cells, or pixels, and calculates the number of raypaths that intersect each cell. The result is a matrix of simultaneous velocity equations with a non-unique solution. The analysis then diverges from an initial estimated model of velocity structure (i.e., horizontal layering) to find a 'best fit' velocity with the observed data, performing multiple iterations and adjusting the model.   The more raypaths or "hits" for each cell, the better definition of transmission properties. Fewer raypaths provide a less certain solution. For this reason, data quality is often low in the corners and edges of tomographic images, and we use only the high quality data from the center. For processing, all the COP and

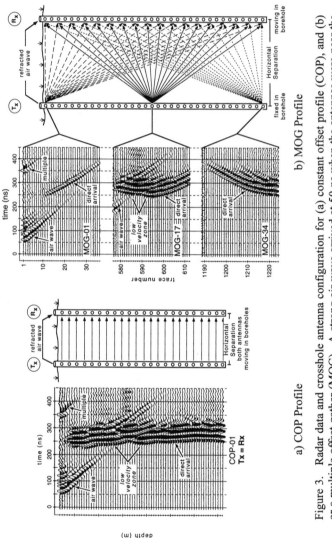

a) COP Profile

b) MOG Profile

Figure 3.   Radar data and crosshole antenna configuration for (a) constant offset profile (COP), and (b) or a multiple offset gather (MOG). A strong air wave arrived at 50 ns when the antennas were near the ground surface, but disappeared at depth. Small circles in each borehole depict individual antenna positions. For clarity, only three sets of raypaths and three data files are shown from a total of 35. Arrival times of some traces in MOG-17 deviate from the expected hyperbola, indicating inhomogeneities in the sediment.

MOG traces were merged together into a single dataset; first arrivals were picked; and a tomographic image was produced that shows variations in velocity (Fig. 4). Each transmitter-receiver path length was used to convert travel-times to velocities within the illuminated plane. The pixel footprint within the resulting radar tomogram contains the averaged velocity from all of the waveforms passing through each pixel space.

RELATION OF SOIL VOID RATIO TO RADAR-WAVE VELOCITY AND DIELECTRIC PROPERTIES

The velocity of radar in a soil mass and its relation to soil-moisture content was developed through empirical studies using the time-domain reflectometry method in a laboratory setting (Topp, et al. 1980). The travel-time of an electromagnetic pulse through a soil mass can be determined through reflection, refraction, or transillumination techniques. At Treasure Island, we used a transillumination approach with crosshole GPR to determine travel-times through a plane in the soil column. Two identical GPR surveys were collected using the same pair of boreholes, one survey prior to blasting and a second survey after blasting. A computer-generated tomogram was computed for each survey, showing the velocity structure of the same section of soil both before (Fig. 4a) and after (Fig. 4b) the blast-induced liquefaction. Multiple offset gathers (MOG) were taken through the soil plane prior to liquefaction, and after blast-induced liquefaction, and a computer generated tomogram was developed for the plane.

The earliest work on the empirical relationship between soil moisture content and radar velocity can be found in Topp et al., 1980. Through the laboratory technique of time-domain reflectometry (TDR), a method essentially identical to our field field approach with crosshole GPR, Topp, et al. (1980) found that a singular relationship exists between the volumetric soil moisture θ_v, (volume of water to the volume of the total soil mass) and the real part of the complex dielectric constant. The complex dielectric constant of soil,

$$\varepsilon = \varepsilon' + j\varepsilon'' \quad [1],$$

is composed of real (ε') and imaginary ($j\varepsilon''$) parts. The radar velocity is dependent only on the real part of the dielectric constant, where $v = c/(\varepsilon')^{1/2}$ and c is the velocity of light in air. The real part of the dielectric constant varies from 1 in air ($v = c = 0.3$ m/nsec) to in excess of 30 for fine grained soils ($v \sim c/6 \sim 0.05$ m/nsec). The radar velocity in soil varies from 0.15 m/nsec in dry sand and 0.06m/nsec for saturated sand, to below 0.05 m/nsec for soft cohesive soil. For a suite of soils types and water contents,

Topp et al. (1980) found the following relation between ε' and θ_v for θ_v between 0 and 0.6,

$$\varepsilon' = 3.03 + 9.3 \bullet \theta_v + 146 \bullet \theta_v^2 - 76.7 \bullet \theta_v^3 \quad [2].$$

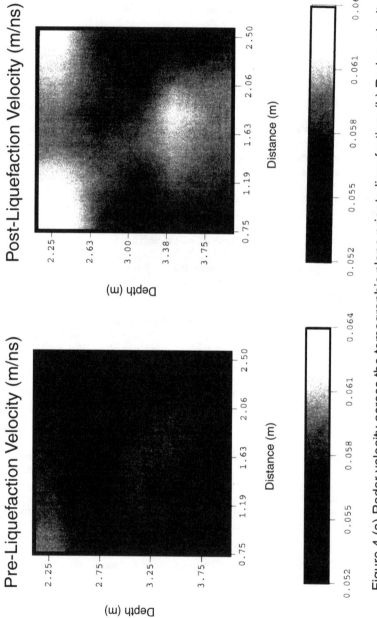

Figure 4 (a) Radar velocity across the tomographic plane prior to liquefaction; (b) Radar velocity after the blast liquefaction event.

Saturated soils have all of the void space filled with water. Under such conditions $\theta = n$, the soil porosity. Resolving equation [2] for T and assuming full saturation, the porosity n can be determined from the real part of the dielectric constant as follows:

$$n = -0.080607 + 0.037649 \cdot \varepsilon' - 0.0011413 \cdot \varepsilon'^2 - 1.5789E\text{-}5 \cdot \varepsilon'^3 \qquad [3].$$

In the field we measure radar velocity, rather than dielectric constant. To estimate porosity directly from radar velocity, the relationship $\varepsilon' = (c/V_r)^2$ (where $c=0.3m/nsec$) is used to modify [3], as follows:

$$n = 2.5025 - 75.54 \cdot V + 920.1 \cdot V^2 - 4094.8 \cdot V^3 \qquad [4].$$

For equation [4], and [6] below, velocity is presented in units of m/nsec. The geotechnical characterization of density state is typically done in terms of void ratio (e). Void ratio is the void volume normalized by the volume of the dry sediment grains. We substituted void ratio for porosity in equations [1] and solved for e in terms of dielectric constant and velocity:

$$e = -0.035129 + 0.030695 \cdot \varepsilon' - 3.5531E\text{-}4 \cdot \varepsilon'^2 + 9.6159E\text{-}6 \cdot \varepsilon'^3 \qquad [5]$$

and

$$e = 13.482 - 533.47 \cdot V + 7526.4 \cdot V^2 - 36615 \cdot V^3 \qquad [6].$$

RESULTS: DENSITY CHANGE DURING LIQUEFACTION

The equations [4] and [6], above, were used to map porosity and void ratio in the soil prior to, and following, blast-induced soil liquefaction. The velocity and density tomograms displayed in this section are for the largest plane in the 3-4-5 m triangle. To better assess density changes in the hydraulic fill, the degraded waveforms passing through the Bay Mud were truncated from the dataset so that an initial homogeneous velocity model could be used. Including the Bay mud in the data set added poorly constrained velocities. The analysis of changes in density prior to and following liquefaction are made on the central portion of the tomogram, a plane extending from 2 to 4 meters in depth and 0.75 to 2.6 meters in width.

Prior to liquefaction, the plane was illuminated by GPR and a velocity tomogram was constructed. Equation 6 was used to convert the image to a pre-liquefaction void ratio tomogram. The radar velocity in the central portion of the soil column at Treasure Island ranged between 0.054 and 0.6 m/ns, and an average velocity of 0.057 m/ns. These velocities translate into void ratios ranging from 0.846 to 0.647, and an average void ratio of 0.738 (Figure 5a). Generally, there is a zone of low to intermediate void ratios in the central portion of the image and a low void ratio zone in the upper left region. A locally high void ratio on the right side of the plane is seen at depths of 2.5-2.75 meters. A higher void ratio zone also is found in the lower left corner of the image.

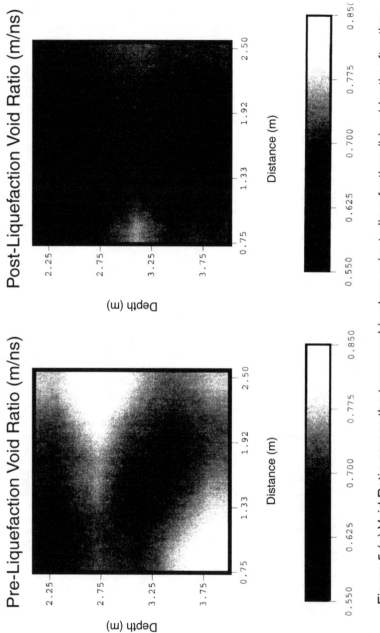

Figure 5 (a) Void Ratio across the tomographic plane prior to liquefaction; (b) void ratio after the blast liquefaction event.

The blasting event occurred in January of 1999, liquefying the hydraulic fill at the site. Elevated pore water pressures were measured in transducer arrays, sand boils were observed at the site and settlements were recorded. Extensive sand boiling and water flow to the surface were observed adjacent to the radar borehole array, such that we are confident that liquefaction occurred within the tomographic plane.

Following the liquefaction event, we resurveyed the 3-4-5 m triangular borehole array and found the velocities had risen considerably throughout the tomogram. The velocity range for the post-liquefaction soil plane ranged from 0.056 to 0.064 m/ns, and an average velocity of 0.0597 m/ns (Figure 4b). The void ratios associated with these velocities range from 0.554 to 0.770, and an average of 0.664 (Figure 5b). By comparing Figures 5a and 5b it can be seen that almost the entire tomographic plane underwent some level of densification and reduction of void ratio during liquefaction.

Detailed imagery of the pre- and post-liquefaction void ratios allow for the differencing of the two tomograms so that we might see where changes in void ratio occurred within the soil column. Figure 6a presents a difference tomogram that shows the initial pre-liquefaction void ratios minus the post-liquefaction values. On average the image shows a densification (reduction in void space) of $\Delta e = 0.074$. The range of void ratio change spans from -0.066 to 0.172. That is, the entirety of the tomographic plane densified, with the exception of a narrow zone at 3.1 meters on the left side of the tomogram. That zone apparently loosened during the liquefaction event, or formed a void when sand was ejected to the surface. Void ratio change, expressed as a percent difference from the initial void ratio state, is presented in Figure 6b. This percent is taken as the values used in Figure 7 divided by the initial void ratio, and so the regions of void ratio change in Figures 6a and 6b look somewhat similar. The average volumetric strain due to void ratio reduction was 4.2 %. Given that the estimated thickness of the liquefied layer at the site was 4 meters, this strain would result in 17.0 cm (6.7 inches) of settlement.

The observed surface settlement at the pile test site can be used as an independent measure of the volumetric strain. The ground level of the site was measured prior to the blasting and afterwards when the pore water pressures had dissipated. Maximum settlements of 16.8, 17.8, and 20.7 cm (6.6-8.2 inches) were recorded at the 4-pile group along three transects across the site measured by BYU engineers. The settlements estimated from the volumetric strain recorded in the radar tomograms are remarkably similar to the observed settlement. Volumetric strains of similar magnitude are reported using resistivity measurements by Arulanandan and Sybico (1993) on sands liquefied during centrifuge models tests conducted at the University of California at Davis.

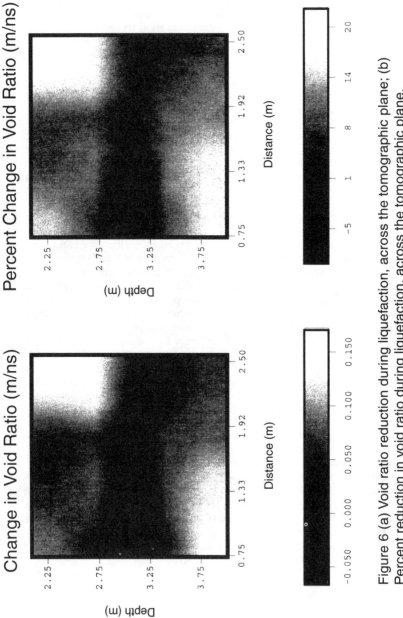

Figure 6 (a) Void ratio reduction during liquefaction, across the tomographic plane; (b) Percent reduction in void ratio during liquefaction, across the tomographic plane.

CONCLUSIONS

We found that the imagery developed through crosshole GPR surveys were able to quantify in spatial detail the initial density state of the soil, and the changes in density due to a liquefaction event. The volumetric strain associated with the estimated changes in density from the radar data would result in settlements of 17cm for the 4 meter thick liquefied layer estimated for the site. The radar-based estimates of settlement are remarkably close to ground level changes observed at the site.

If the critical state void ratio-effective stress line of the soil mass is known, a measure of the height above the steady state line can be presented in tomographic form. That height, known as Ψ (Been and Jefferies, 1985) is useful for assessing flow-failure potential of soil. Likewise, if the minimum and maximum densities were known, a relative density tomogram could be made to assess liquefaction susceptibility. For the Treasure Island site, we intend to collect these additional soil properties so that we may explore the usefulness of GPR for liquefaction assessment. Nevertheless, this paper demonstrates the applicability of cross hole GPR for non-destructive imaging of the spatial structure of void ratio within a large volume of soil, and to characterize changes in soil volume due to a liquefaction event.

REFERENCES

Arulanandan, K. and Sybico, J. Jr. (1993) Post-Liquefaction Settlement of Sands, in Predictive Soil Mechanics, Thomas Telford Publishers, London UK.

Been, K. and Jeffries, M.G. (1985) A state parameter for sands, Geotechnique, 35(2) p.99-112.

Bennett, M. (1994) "Subsurface Investigation for Liquefaction Analysis and Piezometer Calibration at Treasure Island Naval Station, California," Open File Report 94-709, U.S. Geological Survey, 41 p.

Dobry, R., Abdoun, T., and O'Rourke, T.D. (1996). "Evaluation of Pile Response Due to Liquefaction Induced Lateral Spreading of the Ground," Proceedings of the Fourth Caltrans Seismic Design Workshop, 10 p.

Narin van Court, W.A. and Mitchell, J.K. (1995). "New Insights into Explosive Compaction of Loose, Saturated, Cohesionless Soils," Soil Improvement for Earthquake Hazard Mitigation, Geotechnical Special Pub. No. 49, p. 51-65.

Studer, J. and Kok, L. (1980). "Blast Induced Excess Porewater Pressure and Liquefaction Experience and Application," Proceedings of the Intl. Symp. On Soils under Cyclic and Transient Loading, Swansea, Wales, p. 581-593.

Topp, G.C. Davis, J.L., and Annan, A.P. (1980) Electromagnetic Determination of Soil Water Content: Measurement in Coaxial Transmission Lines, Water Resources Research, Vol. 16, No. 3, pp.574-582.

Wilson, D.W., Boulanger, R.W., Kutter, B.L., and Abghari, A. (1996). "Soil-Pile-Superstructure Interaction Experiments with Liquefiable Sand in the Centrifuge," Proceedings of the Fourth Caltrans Seismic Design Workshop, 12 p.

Laboratory Correlation of Liquefaction Resistance with Shear Wave Velocity

Alan F. Rauch,[1] M. ASCE, Michael Duffy[2], A. M. ASCE,
and Kenneth H. Stokoe, II[3], M. ASCE

Abstract

Laboratory data, which relate the liquefaction resistance of two sandy soils to shear wave velocity, are presented and compared to liquefaction criteria derived from seismic field measurements. Recent studies using field case history data have lead to new criteria for assessing liquefaction potential in saturated, granular deposits based on in situ, stress-corrected shear wave velocity. However, this approach is hindered by the relatively small number of case histories and the limited range of site conditions represented in the data catalog. Additional data are needed to more reliably define liquefaction resistance as a function of shear wave velocity. Because shear wave velocity can be measured in situ and in the laboratory, laboratory testing can be used to augment the available field data. In the work described herein, cyclic triaxial and resonant column tests were conducted on specimens of a clean uniform sand and a silty sand. Cyclic undrained strength and small-strain shear wave velocity were determined for identical specimens formed by water sedimentation. The data from these tests were found to be consistent with published field performance criteria, even with the uncertainties of relating laboratory data to field response. This study demonstrates the link between field and laboratory measurements that is possible with shear wave velocities. This link creates the opportunity to extend this approach to study other materials, such as silty sands and gravelly soils, and to study the influence of other parameters, such as high confining pressure, where little to no field performance data are available.

[1] Assistant Professor, Dept. of Civil Engineering, The University of Texas at Austin, Austin, TX 78712.

[2] Staff Engineer, Langan Engineering and Environmental Services, New York, NY 10006

[3] Professor, Dept. of Civil Engineering, Dept. of Civil Engineering, The University of Texas at Austin, Austin, TX 78712.

Introduction

In engineering practice, the liquefaction potential of saturated, granular soils is usually evaluated using empirical correlations with in situ penetration resistance. The "simplified procedure" proposed by Seed and Idriss (1971) can be used to evaluate liquefaction resistance based on standard penetration test (SPT) blowcounts. Seed and Idriss originally developed this procedure using field performance data from sites shaken by earthquakes, coupled with a basic understanding gained from laboratory tests on liquefiable soils. Over the years, the simplified procedure has been modified and updated with additional data, and has become the most commonly used way to assess the potential for granular soils to liquefy. Similar methods have been developed to evaluate liquefaction resistance from cone penetration test (CPT) data (Robertson and Wride 1997) and Becker penetration test data (Harder 1997). Recent studies have led to a new technique for assessing potential liquefaction based on stress-corrected shear wave velocity (Andrus et al. 1997; 1999).

While liquefaction criteria based on SPT and CPT data are fairly well developed, penetration tests may be undesirable, impractical, or unreliable at some sites. Because shear wave velocity (V_S) can be measured more economically and reliably in some situations, a method for evaluating liquefaction resistance based on V_S provides an attractive alternative and/or supplementary means of assessment. The advantages of a shear wave velocity-based method have been pointed out by Dobry et al. (1981), Stokoe et al. (1988), Tokimatsu and Uchida (1990), Finn (1991), Robertson et al. (1995), and Andrus and Stokoe (1997):

- Measurements may be made in soils where penetration data are difficult to obtain or unreliable, such as gravelly soils.
- Shear wave velocity can be measured at sites where intrusive methods are not permitted or where access for drilling equipment is limited, such as landfills, earthfill dams, underwater, and beneath existing structures.
- In situ shear wave velocity is a soil property with a clear physical meaning and is used for many dynamic analyses.
- The necessary measurements are made without disturbing the in situ soil deposit, eliminating concerns associated with sample disturbance.

The development of correlations between V_S and liquefaction resistance has been hampered by the relatively small database of field case histories that is currently available. However, an additional advantage of using V_S is that shear wave velocity can be measured in both laboratory samples and in field deposits. Hence, laboratory studies can be used, in conjunction with field performance data, to broaden the applicability of liquefaction criteria based on V_S.

In this study, cyclic triaxial and resonant column tests were conducted on reconstituted specimens of two sandy soils. In this way, liquefaction resistance and shear wave velocity were measured in identical laboratory samples and then compared to the field performance curves proposed by Andrus et al. (1999). The limited data generated in this study are not sufficient to justify modifying field correlations between

V_S and liquefaction resistance. However, this approach could be used, with a more extensive set of laboratory data, to guide construction of improved liquefaction criteria from field performance data.

Correlations between Shear Wave Velocity and Liquefaction Resistance

A soil is predicted to liquefy if the cyclic stresses induced by an earthquake (represented by the *cyclic stress ratio* or *CSR*) exceed the undrained cyclic strength (represented by the *cyclic resistance ratio* or *CRR*) of the deposit (Robertson and Wride 1997). The factor of safety against triggering liquefaction can be expressed as the ratio of *CRR/CSR*. The *CSR* generated by an earthquake, as well as the *CRR* of a soil deposit, are computed for the given earthquake magnitude or representative number of load cycles as the average cyclic shear stress divided by the initial vertical effective stress. To evaluate the potential for liquefaction at a site based on shear wave velocity (V_S), a correlation between *CRR* and V_S is needed.

Correlations from laboratory data. The relationship between liquefaction resistance and shear wave velocity has been studied in the laboratory by De Alba et al. (1984) and Tokimatsu and his colleagues (1986; 1990; 1991). For a soil liquefied in a cyclic triaxial test, the imposed cyclic stress ratio is defined as the maximum shear stress in a load cycle, τ_{max}, divided by the initial effective confining pressure, σ'_c. That is, $CSR = \tau_{max}/\sigma'_c = \sigma_{dc}/2\sigma'_c$, where σ_{dc} is the cyclic deviator stress. The cyclic resistance ratio measured in a triaxial test (CRR_{triax}) can be defined as the cyclic stress ratio causing liquefaction in 15 uniform cycles of load, where "liquefaction" is defined as the point where 5% double amplitude axial strain occurs (Ishihara 1996). Fifteen load cycles are appropriate for a $M_W = 7.5$ earthquake (Seed et al. 1983); magnitude scaling factors can be used to adjust *CRR* to represent other earthquake magnitudes (Youd and Noble 1997).

Cyclic triaxial test data obtained by De Alba et al. (1984) and Tokimatsu et al. (1986; 1990) indicated a non-unique relationship between V_S and *CRR* for different soils. However, Tokimatsu and Uchida (1990) accounted for the effects of soil type and confining pressure by normalizing their data with respect to minimum void ratio and mean effective stress. Further, because the minimum void ratio is largely determined by soil gradation, they suggested using fines content in the liquefaction evaluation procedure to correct for different soil types. Subsequently, Tokimatsu et al. (1991) presented a laboratory correlation between cyclic triaxial strength and V_S normalized with respect to mean effective stress. They presented two correlation charts, one for clean sands and one for silty sands. Andrus and Stokoe (1997) compared these charts with measured field data and found that the procedure was nonconservative when the number of load cycles was larger than 10 and the stress-corrected V_S was greater than 150 m/s.

Correlations from field performance data. To develop a correlation between *CRR* and V_S directly from observed field performance, one needs a catalog of case history sites where liquefaction was or was not observed following historic earthquakes. Values of the *CSR* induced by the earthquake, normalized with respect to magnitude, are plotted against the in situ V_S. A bounding curve is then developed to divide sites where liquefaction was observed from those sites where no liquefaction was evident. Then, for a given V_S, this bounding curve can be used to estimate the *CRR* of a soil deposit. That is, liquefaction would be expected if a seismic event generates a *CSR* greater than the *CRR* predicted from V_S and the bounding curve, for the given earthquake magnitude.

Liquefaction assessment criteria of this type have been developed by Stokoe et al. (1988), Robertson et al. (1992), Kayen et al. (1992), Lodge (1994), and Andrus and Stokoe (1997). Andrus et al. (1999) compiled the most comprehensive field performance database, which includes 70 sites and 20 earthquakes. The case history data, scaled to represent a $M_W = 7.5$ event, are plotted in Figure 1. Based on this data, Andrus et al. (1999) recommend the following equation for estimating the cyclic resistance ratio (*CRR*) of an in situ soil deposit from shear wave velocity:

$$CRR = \left\{ 0.022 \left(\frac{CV_{S1}}{100} \right)^2 + 2.8 \left(\frac{1}{V_{S1}^* - CV_{S1}} - \frac{1}{V_{S1}^*} \right) \right\} \left(\frac{M_W}{7.5} \right)^{-2.56} \tag{1}$$

where M_w is the moment magnitude of the earthquake. V_{S1} is the measured shear wave velocity (V_S) normalized with respect to the effective vertical stress (σ'_v):

$$V_{S1} = V_S \left(P_a / \sigma'_v \right)^{0.25} \tag{2}$$

where P_a is 100 kPa or approximately one atmosphere. In Equation 1, C is a correction factor meant to account for high values of V_{S1} that result from cementation, aging, or negative pore water pressures ($C \le 1$). The parameter V_{S1}^* is a critical value of V_{S1} that separates contractive and dilative behavior at large strains and is assumed to vary with fines content (*FC*):

$$V_{S1}^* = 215 \text{ m/s} \qquad \text{for } FC \le 5\% \tag{3a}$$
$$V_{S1}^* = 215 - 0.5(FC-5) \text{ m/s} \qquad \text{for } 5\% < FC < 35\% \tag{3b}$$
$$V_{S1}^* = 200 \text{ m/s} \qquad \text{for } FC \ge 35\% \tag{3c}$$

Equations 1 through 3 were used to draw the curves shown in Figure 1, which conservatively bound more than 95% of the liquefaction events in the database.

Presently, the application of these techniques is hindered by the limited range of conditions represented within the field performance database. For example, the field data available to support the definition of V_{S1}^* in Equation 3 are relatively sparse. The correction factor C is also poorly defined. Data from more case histories are needed, particularly for deep deposits (greater than 8 m), dense soils (V_S greater than 200 m/s), and strong ground motions (peak horizontal acceleration greater than 0.4 g) (Andrus et al. 1999). However, in the absence of more field performance data, laboratory tests can be used to improve and extend the correlation between *CRR* and V_S.

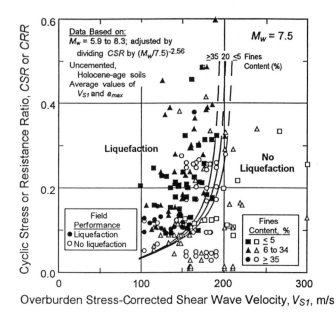

Figure 1. Curves for predicting liquefaction resistance from shear wave velocity, computed using Equations 1-3, shown with case history data (from Andrus et al. 1999).

Soils Tested in This Study

Tests were conducted on two soils: a poorly graded uniform sand and a silty sand. Identical specimens of the different soils, which were reconstituted by sedimentation in water, were tested in a cyclic triaxial apparatus and a resonant column device. The cyclic triaxial apparatus was used to measure the cyclic strengths of the silty sand at one density and the uniform sand at two densities. The resonant column device was used to measure the shear wave velocities of the samples at small strain levels, similar to the strain levels used in field seismic measurements.

The first soil tested, Monterey #0/30 sand, is a uniformly graded, subrounded, medium-grained sand (SP). It is a commercially produced, washed and sieved beach sand. Monterey #0/30 sand was chosen so that test results could be compared with data previously published by others. The second soil, identified here as Wyoming sand, is a fine-grained, subangular, silty sand with a trace of fine gravel (SM). Table 1 summarizes the measured index properties of the two soils, and the measured grain size distributions are shown in Figure 2.

Table 1. Measured index properties of tested soils.

	Monterey #0/30 Sand	Wyoming Sand
USCS Classification	SP	SM
Percent Finer (%) than 0.075 mm	0.0	13
Specific Gravity of Solids, G_s	2.62	2.66
Maximum Dry Unit Weight, $\gamma_{d,max}$ (pcf)	105.1	109
Minimum Dry Unit Weight, $\gamma_{d,min}$ (pcf)	89.8	90
Minimum Void Ratio, e_{min}	0.56	0.52
Maximum Void Ratio, e_{max}	0.82	0.85
Atterberg Limits	--	nonplastic

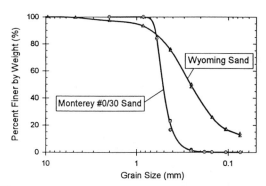

Figure 2. Measured grain size distribution of tested soils.

Specimen Preparation Procedures

All samples in this study were formed by a water sedimentation technique, which was chosen in an attempt to recreate the soil fabric produced in a natural depositional process. The test specimens were 70 mm in diameter, with a height-to-diameter ratio between 2.1 and 2.4. A pre-weighed amount of air-dry soil was spooned into a forming mold filled with de-aired, de-ionized water. The water level was kept above the soil surface at all times, and the soil was added slowly to minimize the segregation of fines. Once the prescribed amount of soil was in place, a surcharge (1.3 kg of steel) was placed on the soil and the mold was tapped until the soil was compacted to the desired density. This same procedure was used to form specimens for both cyclic triaxial and resonant column testing.

A pore pressure of at least 500 kPa was used to backpressure saturate the cyclic triaxial specimens. Saturation was verified by measuring Skempton's pore pressure parameter; nearly all test specimens exhibited $B \geq 0.98$. An isotropic state of stress was maintained at all times during backpressure saturation and subsequent consolidation of the test specimens. All samples were consolidated to an effective isotropic stress of 100 kPa. Sample deformations during saturation and consolidation were measured and used to determine the final sample density.

One specimen of Wyoming sand and three specimens of Monterey #0/30 sand were tested in the resonant column apparatus. These samples were prepared in the exact same manner as those for the triaxial tests, except that the resonant column specimens were not backpressure saturated. However, because they were formed by sedimentation in de-aired water, the resonant column specimens are assumed to have been near full saturation. Since only small-strain measurements were performed during resonant column testing, degrees of saturation slightly less than 100% are acceptable and the exact value is inconsequential.

Measured Cyclic Triaxial Strengths

The triaxial specimens were sheared under a sinusoidal axial load applied at a frequency of 0.25 Hz. A loading frequency less than the preferred rate of 1 Hz was chosen to allow optimal control of the loading function. Axial loads were measured using a load cell inside the triaxial cell, to eliminate errors due to piston friction. Following Ishihara (1996), liquefaction was defined as the first occurrence of 5% double amplitude axial strain.

The measured cyclic triaxial strength of medium dense Monterey #0/30 sand, in terms of the number of cycles to cause liquefaction at a given *CSR*, is indicated in Figure 3. Test data from this study fall within the middle of the scatter of previously

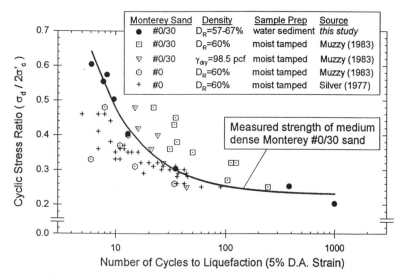

Figure 3. Cyclic triaxial strength of medium dense Monterey sand.

published data on Monterey sand. However, considering differences in gradation, sample preparation, and the definition of liquefaction failure, the strengths measured in this study are consistent with the data reported by others. For example, the data reported by Silver (1977) were from tests on Monterey #0 sand, which is no longer produced. Muzzy (1983) tested both Monterey #0 and #0/30 sands and showed that, for the same relative density, Monterey #0/30 is somewhat stronger (more cycles to liquefy under a given cyclic stress ratio). Consistent with Muzzy's findings, the results from this study plot above the published data for Monterey #0. In addition, Mulilus et al. (1977) showed that samples formed by water sedimentation (used in this study) are somewhat weaker than samples formed by moist tamping (used in the other tests plotted in Figure 3). Accordingly, the data from this study plot in Figure 3 on the lower bound of the published data for moist-tamped Monterey #0/30 sand. Finally, the previously published test data represent the number of load cycles to cause 10% double amplitude axial strain, whereas the test results from this study are derived from defining liquefaction as the first occurrence of 5% double amplitude strain. Considering all of these factors, the cyclic triaxial data obtained in this study are found to be consistent with previously published data on Monterey sand.

The measured cyclic triaxial strengths of Wyoming sand ($D_r \cong 15\%$) and Monterey #0/30 sand ($D_r \cong 20\%$ and 60%) are plotted in Figure 4. Some of the scatter in the measured strengths can be attributed to small differences in specimen densities, but the measured strengths plotted in this figure were not adjusted for this effect. The cyclic triaxial strengths (CRR_{triax}) reported in Table 2 were determined from the curves

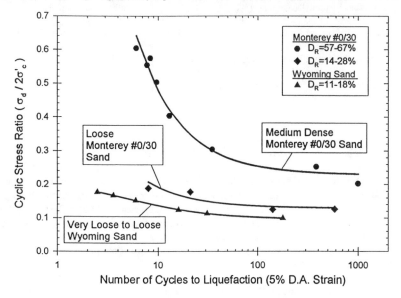

Figure 4. Measured cyclic triaxial strengths of tested soils.

in Figure 4 at 15 load cycles, which is assumed to represent a $M_W = 7.5$ earthquake as discussed earlier.

In applying the results from cyclic triaxial compression tests, the CRR should be adjusted to account for conditions in the laboratory test that are different from the field. In an approximate manner, this correction can be made using (Seed 1979; Tokimatsu and Uchida 1990):

$$CRR_{field} = (1/3)(1 + 2K_0) r_c CRR_{triax} \qquad (4)$$

where r_c is a constant that accounts for the effect of multidirectional shaking (value of 0.9 to 1.0) and K_0 is the lateral earth pressure coefficient. Equation 4 is used to compensate for the effects of testing isotropically consolidated specimens along a triaxial stress path, which differs from the earthquake loading of in situ soils. A value of $K_0 = 0.47$, representative of a sand with $\phi' = 32°$, was assumed for both soils in this study. Assuming $r_c = 0.95$, the estimated field cyclic strengths are then given by the relationship: $CRR_{field} = (0.61)CRR_{triax}$. The liquefaction resistance of Monterey #0/30 sand and Wyoming sand, determined in this manner, are given in Table 2.

Table 2. Measured liquefaction resistance and shear wave velocities.

	Monterey #0/30 Sand		Wyoming Sand
Approx. Relative Density, D_R (%)	20	60	15
Void Ratio for D_R	0.77	0.66	0.80
Cyclic Resistance Ratio for triaxial conditions, CRR_{triax}	0.17	0.40	0.13
Cyclic Resistance Ratio for field conditions, CRR_{field}	0.10	0.24	0.08
Stress-Corrected Shear Wave Velocity, V_{S1} (m/s)	186	199	150

Shear Wave Velocities from Resonant Column Tests

Small-strain shear wave velocities (V_S) and shear moduli (G_{max}) of the two soils were determined from tests in a torsional resonant column device, with specimens consolidated under three to four different confining pressures. As described by Stokoe et al. (1994), resonant frequencies and amplitudes of vibration measured in this test, along with a system calibration, are used to determine V_S. The corresponding values of the stress-corrected shear wave velocity (V_{S1}) were determined in a two-step procedure as described below.

First, the values of G_{max} measured for each soil were used to fit the empirical relationship in Equation 5 (Hardin 1978):

$$G_{max} = \frac{A}{0.3 + 0.7e^2} \sigma'_m{}^n OCR^k P_a^{1-n} \tag{5}$$

where e is void ratio, σ'_m is mean effective stress, OCR is the overconsolidation ratio (OCR = 1 in these tests), and P_a is atmospheric pressure. The dimensionless stiffness parameter (A) and the pressure exponent (n) are empirical coefficients that are determined from the measured values of G_{max}. The best fit to the resonant column test data on Wyoming sand gave $A = 411$ and $n = 0.586$. For Monterey #0/30 sand, the best fit gave $A = 585$ and $n = 0.515$.

Next, a stress-corrected shear wave velocity, corresponding to that defined in Equation 2, was computed for each soil at the appropriate density. In the field, V_{S1} would be determined as the velocity of a shear wave in which the direction of particle motion and the direction of wave propagation are polarized along horizontal and vertical principal stress directions, in soil under one atmosphere of vertical effective stress (Stokoe et al. 1999). Given the values of A and n determined above, the appropriate values of V_{S1} were then computed from:

$$V_{S1} = \sqrt{\frac{G_{max}}{\rho}} = \sqrt{\frac{A}{(0.3 + 0.7e^2)\rho} \left(\frac{1 + 2K_0}{3}\right)^n \sigma'_v{}^n P_a^{1-n}} \tag{6}$$

where ρ is the mass density of the soil with a void ratio of e, and K_0 is assumed to be 0.47 for both soils as before. The resulting values of V_{S1} are shown in Table 2.

Comparison of Laboratory Results to Field Correlation

The data in Table 2, which correlate V_{S1} to CRR for Monterey #0/30 and Wyoming sands, are plotted in Figure 5 together with boundary curves for $M_W = 7.5$ developed by Andrus et al. (1999) from field performance data (Equations 1 through 3). Because adjusting CRR_{triax} to get CRR_{field} based on Equation 4 is only approximate, both the measured cyclic strengths (CRR_{triax}) and the corresponding, adjusted field values (CRR_{field}) are plotted in Figure 5. Solid and open symbols are used to represent CRR_{triax} and CRR_{field}, respectively, in this plot.

As can be seen in Figure 5, the laboratory data from this study follow the same trends as the field performance curves; that is, increasing liquefaction resistance with increasing relative density, shear wave velocity, and fines content. If all of the laboratory tests had been in exact agreement with the field performance criteria, the values of CRR_{field} (open symbols) would plot on top of the boundary curves in Figure 5. Indeed, results from tests on two of these three materials are in excellent agreement with the field performance curves. However, the third data point, for the loose Monterey sand, plots well below the boundary curve. This deviation at a V_{S1} of 186 m/s suggests the need for more tests, to determine if this data is anomalous or if the shape of the boundary curve needs to be refined in this regime. In any case, these results indicate that much more testing on a range of soils would be beneficial.

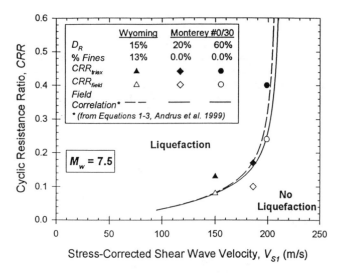

Figure 5. Comparison of laboratory data with field performance criteria for a M_W = 7.5 earthquake.

The curves in Figure 5 also illustrate where liquefaction evaluations, based on the method proposed by Andrus et al. (1999), are sensitive to the precision of the measured shear wave velocity. For values of V_{S1} in the range of about 170 to 220 m/s, where the curves in Figure 5 turn sharply upward, small changes in V_{S1} correspond to large changes in CRR. This implies the need for precise measurements of shear wave velocity to accurately assess liquefaction resistance when V_{S1} is in this range. However, for V_{S1} less than about 170 m/s, the change in CRR with V_{S1} is more gradual. In this lower range of material stiffness, there is a strong possibility of liquefaction when the soil is subjected to even relatively small cyclic shear stresses. On the other hand, values of V_{S1} above 220 m/s indicate that liquefaction is very unlikely, for the soil types represented in the database, regardless of the level of shaking.

Another way to compare laboratory data with the field performance criteria is shown in Figure 6. In this plot, the data points and trend lines from Figure 4 are reproduced using solid symbols and dotted lines, respectively. The solid lines in Figure 6 show the corresponding undrained cyclic strength in triaxial loading as predicted with the field performance criteria developed by Andrus et al. (1999). These predictions have been adjusted using Equation 4 to yield CRR_{triax}. As an example, consider the predicted cyclic triaxial strength of Wyoming sand that is obtained from Equation 1 using V_{S1} = 150 m/s (Table 2), V_{S1}^* = 211 m/s (from Equation 3b for 13% fines), and C = 1 (uncemented, reconstituted sample with positive pore pressures). For M_W = 7.5, this yields CRR_{field} = 0.082; applying the correction given by Equation 4 gives CRR_{triax} = $CRR_{field}/0.61$ = 0.135. This value is then plotted at 15 cycles of

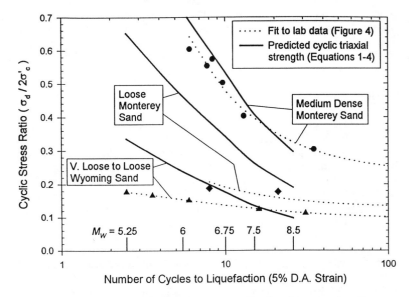

Figure 6. Comparison between predicted and measured cyclic triaxial strengths of tested soils.

uniform load, representative of a $M_W = 7.5$ earthquake. Values for $M_W = 5.25$, 6, 6.75, 7.5, and 8.5 are plotted at 2.5, 5.5, 10, 15, and 26 load cycles, respectively, based on the recommendations of Seed et al. (1983).

For 2 to 26 load cycles (representing earthquakes of $M_W = 5.25$ to 8.5), Figure 6 indicates very good agreement between the laboratory data and the predicted cyclic triaxial strengths for the medium dense Monterey sand. The same comparison for the Wyoming sand is not as good, especially when there is less than about 10 load cycles. The comparison is very poor for the loose Monterey sand, but this may be due to the lack of test data on samples that liquefied in less than 20 load cycles. In general, some of the discrepancies in Figure 6 probably result from the different methods used to represent different earthquake magnitudes ("equivalent" number of uniform load cycles for the laboratory data as opposed to the use of the magnitude scaling factor equal to $(M_W/7.5)^{-2.56}$ in the field performance criteria). More work is needed to substantiate the equivalency of using these two approaches to account for the effect of earthquake magnitude on cyclic strength.

Finally, a potentially significant source of error in this study results from the measurement of cyclic strength and V_S using separate specimens in two different testing apparatuses. This problem could be overcome by using some means to directly measure V_S and cyclic strength in the same test specimen. One way to do this would be to mount piezoelectric bender elements in the top and bottom platens of the triaxial device, as was done by De Alba et al. (1984). Although used successfully for testing loose sand specimens, this approach may yield misleadingly high values of small-strain stiffness for dense specimens. This potential limitation in the use of bender elements, which may result from the formation of preferred stress-wave travel paths, warrants further investigation.

Conclusions

Laboratory data that relate liquefaction resistance (*CRR*) to shear wave velocity (*Vs*) have been presented. The trends in the laboratory data were found to be consistent with the liquefaction boundary curves developed by Andrus et al. (1999) from field performance data. Given a number of uncertainties, as listed below, there is remarkable agreement between the laboratory test data and the field performance criteria for two of the three materials tested. It would be very beneficial to have more data of this kind from laboratory tests on a wide variety of soils to better investigate the liquefaction criteria developed from the field performance data.

Correlations of cyclic triaxial strengths to field performance must consider a number of inherent uncertainties in the interpretation of laboratory results. These uncertainties include:

- Techniques used for forming test specimens in the laboratory significantly affect the measured cyclic strength.
- The cyclic stress path generated by uniform cycles of axial stress in a triaxial test only approximately models an earthquake loading on a soil deposit.
- The criterion used to define the occurrence of liquefaction in a triaxial test does not correspond exactly with the surface manifestation of liquefaction in the field.
- The uncertainties of the relationship between laboratory and field conditions are only approximately accounted for in the correction of cyclic triaxial strengths (*CRR$_{triax}$*) to in situ cyclic resistance ratios (*CRR$_{field}$*).

In this study, the measurement of cyclic strength and shear velocity in separate soil specimens also introduces a potential source of error.

One important point demonstrated in this study is the link between field and laboratory measurements that is possible with shear wave velocities. This point has also been demonstrated in the past by a number of other researchers. Laboratory measurements may be used to study and quantify the effects of various parameters on shear wave velocity and liquefaction resistance. Accordingly, laboratory data could be used to supplement limited field data and develop more reliable correlations based on *Vs*. For example, a larger catalog of laboratory data could be used to better define the shape of the liquefaction boundary curve, the adjustment based on fines content, and the correction for lightly cemented sands. Laboratory data would be particularly useful for constraining the field curves for conditions where field performance data are lacking, such as for denser soils, gravelly deposits, greater depths, and sites subjected to stronger seismic motions. However, given the inherent limitations outlined above, a correlation between in situ liquefaction resistance and shear wave velocity would not be reliable if based solely on laboratory test data. Actual field performance data from past earthquakes and in situ measurements of V_S are clearly needed to develop reliable liquefaction criteria.

Acknowledgement

The authors would like to thank Farn-Yuh Menq and Mehment Darendeli for their assistance in conducting the resonant column tests.

References

Andrus, R. D., and Stokoe, K. H., II, (1997). "Guidelines for evaluation of liquefaction resistance using shear wave velocity." *Proc., NCEER Workshop on Evaluation of Liquefaction Resistance of Soils,* Tech. Report NCEER-97-0022, National Center for Earthquake Engineering Research, T. L. Youd and I. M. Idriss (eds.), December 31, pp. 89-128.

Andrus, R. D., Stokoe, K. H., II, and Chung, R. M. (1999). "Draft guidelines for evaluating liquefaction resistance using shear wave velocity measurements and simplified procedures." *NISTIR 6277,* National Institute of Standards and Technology, Gaithersburg, Maryland.

De Alba, P., Baldwin, K., Janoo, V., Roe, G., and Celikkol, B. (1984). "Elastic-wave velocities and liquefaction potential." *Geotech. Testing J.,* Vol. 7, No. 2, June, pp. 77-87.

Dobry, R., Stokoe, K. H., II, Ladd, R. S., and Youd, T. L. (1981). "Liquefaction susceptibility from S-wave velocity." *Proc., In Situ Tests to Evaluate Liquefaction Susceptibility,* ASCE National Convention, St. Louis, Missouri, October.

Finn, W. D. L. (1991). "Assessment of liquefaction potential and post-liquefaction behavior of earth structures: developments 1981-1991." *Proc., 2nd Int. Conf. on Recent Advances in Geotech. Earthquake Engrg. and Soil Dynamics,* St. Louis, Missouri, March, pp. 1833-1850.

Harder, L. F. (1997). "Application of the Becker penetration test for evaluating the liquefaction potential of gravelly soils." *Proc., NCEER Workshop on Evaluation of Liquefaction Resistance of Soils,* Tech Report NCEER-97-0022, National Center for Earthquake Engineering Research, T. L. Youd and I. M. Idriss (eds.), December 31, pp. 129-148.

Hardin, B. O. (1978). "The nature of stress-strain behavior of soils." *Proc., Earthquake Engineering and Soil Dynamics,* ASCE, Pasadena, Calif., Vol. 1, pp. 3-89.

Ishihara, K. (1993). "Liquefaction and flow failure during earthquakes." Thirty-third Rankine Lecture, *Geotechnique,* Vol. 43, No. 3, pp. 351-415.

Ishihara, K. (1996) *Soil behaviour in earthquake geotechnics.* Oxford Univ. Press, New York.

Kayen, R. E., Mitchell, J. K., Seed, R. B., Lodge, A., Nishio, S., and Coutinho, R. (1992). "Evaluation of SPT-, CPT-, shear wave-based methods for liquefaction potential assessment using Loma Prieta data." *Proc., 4t^h Japan-US Workshop on Earthquake Resistant Design of Lifeline Facilities and Countermeasures for Soil Liquefaction, Tech. Report NCEER-92-0019,* M. Hamada and T. D. O'Rourke (eds.), pp. 177-204.

Lodge, A. L. (1994). "Shear wave velocity measurements for subsurface characterization." Ph.D. Dissertation, University of California at Berkeley.

Mulilis, J. P., Seed, H. B., Chan, C., Mitchell, J. K., and Arulanadan, K. (1977). "Effects of sample preparation on sand liquefaction." *J. Geotech. Engineering Div.*, ASCE, Vol. 103, No. GT2, February, pp. 91-108.

Muzzy, M. W. (1983). "Cyclic triaxial behavior of Monterey No. 0 and No. 0/30 Sands." Master's thesis, Colorado State University, Ft. Collins, Colorado.

Robertson, P. K., and Wride, C. E. (1997). "Cyclic liquefaction and its evaluation based on the SPT and CPT." *Proc. of NCEER Workshop on Evaluation of Liquefaction Resistance of Soils*, Tech Report NCEER-97-0022, National Center for Earthquake Engineering Research, T. L. Youd and I. M. Idriss (eds.), December 31, pp. 41-87.

Robertson, P. K., Woeller, D. J., and Finn, W. D. L. (1992). "Seismic cone penetration test for evaluating liquefaction potential under cyclic loading." *Canadian Geotech. J.*, Vol. 29, No. 4, August, pp. 686-695.

Robertson, P. K., Sasitharan, S., Cunning, J. C., Sego, D. C. (1995). "Shear wave velocity to evaluate in-situ state of Ottawa Sand." *J. Geotech. Engrg.*, ASCE, Vol. 121, No. 3, March, pp. 262-273.

Seed, H. B. (1979). "Soil liquefaction and cyclic mobility for level ground during earthquakes." *J. Geotech. Engrg. Div*, ASCE, Vol. 105, No. GT2, February, pp. 201-255.

Seed, H. B., and Idriss, I. M. (1971). "Simplified procedure for evaluating soil liquefaction potential." *J. Soil Mech. and Found. Div.*, ASCE, Vol. 97, No. SM9, September, pp. 1249-1273.

Seed, H. B., Idriss, I. M., and Arango, I. (1983). "Evaluation of liquefaction potential using field performance data." *J. Geotech. Engrg.*, ASCE, Vol. 109, No. 3, March, pp. 458-482.

Silver, M. L. (1977). "Laboratory triaxial testing procedures to determine the cyclic strength of soils." *Report No. NUREG-31*, U. S. Nuclear Regulatory Commission, Washington, D. C., 129 pages.

Stokoe, K. H., II, Roesset, J. M., Bierschwale, J. G., and Aouad, M. (1988). "Liquefaction potential of sands from shear wave velocity." *Proc., 9th World Conf. on Earthquake Engineering*, Tokyo, Japan, Vol. III, pp. 213-218.

Stokoe, K. H., II, Hwang, S. K., Lee, J. N.-K., and Andrus, R. D. (1994). "Effects of various parameters on the stiffness and damping of soils at small to medium strains." *Proc., Inter. Symposium on Prefailure Deformation Characteristics of Geomaterials*, Vol. 2, Sapporo, Japan, September, pp. 785-816.

Stokoe, K. H., II, Darendeli, M. B., Andrus, R. D., and Brown, L. T. (1999). "Dynamic soil properties: Laboratory, field and correlation studies." *Proc., 2nd Int. Conf. on Earthquake Geotech. Engrg.*, S. Pinto (ed.), pp. 811-845.

Tokimatsu, K., and Uchida, A. (1990). "Correlation between liquefaction resistance and shear wave velocity." *Soils and Foundations*, JSSMFE, Vol. 30, No. 2, June, pp. 33-42.

Tokimatsu, K., Yamazaki, T., and Yoshimi, Y. (1986). "Soil liquefaction evaluation by elastic shear moduli." *Soils and Foundations*, JSSMFE, Vol. 26, No. 1, March, pp. 25-35.

Tokimatsu, K., Kuwayama, S., and Tamura, S. (1991). "Liquefaction potential evaluation based on Rayleigh wave investigation and its comparison with field behavior." *Proc., 2nd International Conf. on Recent Advances in Geotech. Earthquake Engrg. and Soil Dynamics*, S. Prakash (ed.), St Louis, Missouri, March, pp. 357-364.

Youd, T. L., and Noble, S. K. (1997). "Magnitude scaling factors." *Proc., NCEER Workshop on Evaluation of Liquefaction Resistance of Soils*, Tech. Report NCEER-97-0022, National Center for Earthquake Engineering Research, T. L. Youd and I. M. Idriss (eds.), December 31, pp. 149-165.

CONSTITUTIVE MODELING OF FLOW LIQUEFACTION AND CYCLIC MOBILITY

By X. S. Li[1], Member, ASCE, H. Y. Ming[2], and Z. Y. Cai[2]

ABSTRACT

Flow liquefaction is due to contractive response of loose granular soils, while cyclic mobility is related to both contractive and dilative responses of granular soils. These two failure mechanisms may occur in a single soil, depending on the density and confining pressure applied. However, due mainly to the lack of a unified framework to describe the contractive and dilative response of granular soils, these two failure mechanisms are considered separately in current practice.

The key issue in unified treatment of these failure mechanisms is correct modeling of dilatancy. The classical stress dilatancy theory in its exact form is incapable of doing this. The dilatancy must be a function of stress state as well as of the internal material state. Such a state dependent dilatancy in conjunction with the critical state framework can effectively model both flow liquefaction and cyclic mobility over a wide range of material and stress states, using a single set of material constants. This paper describes this unified approach and a model that follows this approach.

INTRODUCTION

A recent trend in evaluating the seismic response and safety is the use of displacement as a criterion for assessing the seismic performance of earth structures and planning remediation measures. This is especially true when soil liquefaction is involved. Post-liquefaction displacement analysis has already been a significant part of geotechnical earthquake engineering practice. The predictive capability of displace-

---

[1] Associate Professor, Department of Civil Engineering, The Hong Kong University of Science and Technology, Clear Water Bay, Kowloon, Hong Kong, China

[2] Graduate Student, Department of Civil Engineering, The Hong Kong University of Science and Technology, Clear Water Bay, Kowloon, Hong Kong, China

ment, however, is largely restricted by the availability of a reliable model that can adequately describe the stress-strain-strength behavior of granular soils.

Liquefaction phenomena can be divided into two groups: flow liquefaction and cyclic mobility. Flow liquefaction is caused by a sudden drop of shear strength and is in connection with the contractive behavior of loose materials, while cyclic mobility is a result of gradually reducing effective confining stress and is associated with both the contractive and dilative responses of granular soils at low confining stresses. These two liquefaction mechanisms are considered separately in the current practice. From the soil mechanics point of view, however, they are tied to a unified framework.

It is well known that subjected to shear, loose granular materials contract and medium-to-dense ones dilate. Whether a material is in a loose state or a dense state depends not only on the density but also on the confining pressure applied. Furthermore, for a material, which initially is either in the loose or dense state, the behavior may change during loading, due to possible changes in the stress and material states. It is thus preferable to have a unified framework to model the state dependent material response, either contractive or dilative, over a full range of stress and material states. With such a framework, the liquefaction phenomena can be described in a unified way.

It has been identified that the key issue in such a unified approach is to properly formulate dilatancy. The classical stress dilatancy theory in its exact form is incapable of doing this (Li and Dafalias 2000). The dilatancy of a granular material must be a function of stress state as well as the internal state of the material. Such a state dependent dilatancy in conjunction with the critical state soil mechanics forms a framework that can effectively model both the contractive and dilative behavior of granular materials over a wide range of material densities and confining pressures, using a single set of material constants.

This paper describes this unified approach and a model that follows this approach. Its capability in unified modeling of flow liquefaction and cyclic mobility is shown.

EXISTING PROCEDURES FOR POST-LIQUEFACTION ANALYSIS

The main steps in current practice for evaluating the seismic response of earth structures containing liquefiable soils include a) determining which soils may liquefy during an earthquake of given motion parameters based on in situ test data; b) determining the residual or steady state strengths of the liquefiable soils; and c) conducting stability analysis and/or post-liquefaction displacement analysis for the earth structure concerned (Finn 1990).

A number of numerical procedures are available for earthquake response analysis (Scott and Arulanandan 1993). Some of them have included the predictive capability for post-liquefaction displacements. For example, TARA-3FL (Finn 1990), a procedure that can be used in practice as well as in research environment, includes a triggering mechanism to switch the strength of any liquefiable elements in the earth dam to the steady state strength at the time a liquefaction triggering criterion is met. This procedure has been used in analyses for Sardis Dam in Mississippi and ground deformation in Niigata City, Japan, during the 1964 earthquake, showing that the calculated displacements match the measured field displacements reasonably well. However, the analyses also indicated that the post-liquefaction deformation was very sensitive to the residual strength adopted and the final deformed shape should always be checked separately by using a conventional stability analysis.

A different procedure was proposed by Gu et al. (1993, 1994) to analyze post-earthquake deformations of lower San Fernando dam and Wildlife site. The stress redistribution analysis was conducted under fully undrained conditions by the finite-element method. In the analyses, a simplified undrained boundary-surface model was used to simulate the behavior of liquefiable soils. A hyperbolic strain-softening model (Chan and Morgenstern 1989) was adopted to simulate the soil behavior during collapse from its peak strength to the steady-state strength. The collapse surface proposed by Sladen et al. (1985) was embedded in the model as a triggering condition for the onset of strain-softening behavior of liquefied materials. As pointed out by Gu et al. (1993), during flow failure, the sand grain structure may collapse, the stored strain energy may release and the load previously carried by the liquefied soils may be transferred to surrounding materials causing further liquefaction or yielding. As a result, the liquefied zone after stress redistribution may be much larger than the initial liquefied zone caused directly by an earthquake.

In the above-cited procedures the response of soil is somehow predetermined. Before-liquefaction and post-liquefaction are treated as two separate processes. In such procedures liquefaction and post-liquefaction deformation are triggered by certain preset criteria. The sensitivity of the triggering criteria to the analysis and the impact of the uncertainties in the evolution of soil states, which the liquefaction triggering depends on, have not been addressed in detail.

UNIFIED APPROACH

Events before and after liquefaction are essentially two divided stages of a single continuous process, and a unified procedure is preferable to treat earthquake response including the post liquefaction deformations. Such a procedure should be theoretically sound and practically feasible. At present, the mainstream along this line is the coupled multiphase continuum framework in conjunction with elastoplastic soil models (possibly enhanced by continuum description of soil microstructures). However, such procedures do not always produce satisfactory results, due mainly to the inadequate simulative capability of the soil models.

A unified procedure needs a unified constitutive model, in particular a model that can reproduce all significant stress-strain responses of granular soils during the entire process of earthquakes. The simulative capability of the model should include the responses of either liquefied soils or the soils not liquefied, and for liquefied soils, the response either before or after liquefaction is triggered. In addition, the response of the model should be loading history dependent, i.e., not predetermined by the analysis.

The essential responses of granular soils to earthquake loading include flow liquefaction and cyclic mobility. These two types of responses pertain to the same material (having the same intrinsic properties) at different states. As both the stress and material states may vary drastically under earthquake loading, a model to simulate the responses over the entire loading history must be based on a comprehensive and consistent constitutive framework.

Consider an earth structure containing the same type of sand. The initial states of the sand throughout the structure are in general different, due to differences in density, confining pressure, and stress ratio. When an earthquake occurs, some parts of the soil may contract while others may dilate. When large deformation occurs after liquefaction is triggered, the configuration of the structure may have changed considerably. As a result, the confining pressures may change accordingly, and due to the generation, redistribution and dissipation of excess pore pressures, the effective confining pressures will have additional change. Furthermore, the densities of the soils in the failure zone may also change significantly. It can be seen that in an earth structure not only the initial stress and material states may vary widely, but the states of a given mass of soils may also change drastically during and after an earthquake; resulting in possible switches between contractive and dilative responses. Therefore, for a unified analysis procedure involving liquefaction, a constitutive model capable of modeling responses for all possible densities under all possible confining pressures is needed. In addition, as partial drainage may occur during post liquefaction deformation, the model should be able to handle drained response as well.

STATE-PARAMETER DEPENDENT SAND MODELS

Attempts have been made in recent years to take into account the state dependent response of sand. The loose or dense state of sand is distinguished in reference to the critical or steady state, at which the effective mean normal stress p', the deviatoric stress q, and volumetric strain ε_v are all constants while the deviatoric strain ε_q continuously develops. Been and Jefferies (1985) introduced a parameter ψ called the state parameter, which measures the difference between the current and critical void ratios at same confining pressure. The definition of the state parameter is illustrated in Fig. 1, when ψ is positive, the material is in a loose state (looser than critical state), when it is negative, the material is in a dense

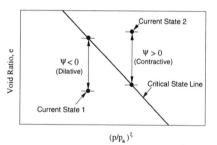

Figure 1. Critical state line and state parameter ψ

state (denser than critical state). When the critical state is reached, $\psi = 0$.

It has been identified (Li and Dafalias 2000) that the key to uniquely relate the contractive and dilative response to stress and material state is the dilatancy $d = d\varepsilon_v^p / d\varepsilon_q^p$, where $d\varepsilon_v^p$ and $d\varepsilon_q^p$ are the increments of plastic volumetric and deviatoric strain, respectively. Manzari and Dafalias (1997) introduced a sand model that defines a linear dependence of the phase transformation stress ratio on ψ, such that when $\psi = 0$ the phase transformation stress ratio becomes identical to the critical stress ratio. The model guarantees satisfaction of the basic premises of critical state soil mechanics, and allows modeling of both loose and dense sand behavior with a unique set of parameters. Li (1997) investigated the response of sand at the ultimate stress ratio and explicitly pointed out that the dilatancy is related not only to the stress ratio but is also a function of plastic volumetric strain. In order to model dilative hardening correctly, care must be taken to distinguish the ultimate stress ratio from the ultimate material state (critical state), at which the dilatancy is equal to zero. The dependence of dilatancy on density as proposed by Li for the ultimate stress ratio, is generalized to all stress ratios (Li et al. 1999) by using the technique proposed by Manzari and Dafalias. A comparison between the simulations by a bounding surface hypoplasticity sand model (Wang et al. 1990) modified to include this ψ dependent dilatancy and test results over a wide range of densities and loading conditions shows that the concept of state-dependent dilatancy works effectively. More recently, Wan and Guo (1998) proposed a model with its dilatancy modified from Rowe's stress-dilatancy equation. The modified dilatancy equation includes a density dependency with the critical void ratio as the reference. Cubrinovski and Ishihara (1998) showed a dilatancy relation that depends on the material state represented by plastic shear strain. Li and Dafalias (2000) discussed a number of issues on the dilatancy, pointed out that a dilatancy function without material state dependence is the major obstacle to unified modeling of the behavior of granular materials over a full range of densities and stress levels.

UNDRAINED RESPONSE OF SAND

Fig. 2 shows typical undrained shear responses of sand with different densities but the same initial confining pressure. Specimen A, which has a low relative density, is initially in a loose state ($\psi > 0$). The sand tends to contract when sheared, resulting in a reduction of the effective confining pressure. When the sand is sheared beyond the point of peak strength, the undrained strength drops to a value that is maintained constant over a large range of shear strain, identified as the steady state deformation (Castro 1969). The strength at that steady state is called the residual strength. This process shows the mechanism of flow liquefaction. Specimen B with a higher relative density is initially in a dense state ($\psi < 0$). The sand tends to dilate after a segment of contractive response at low stress ratios. The reversal from contractive to dilative response occurs at the phase transformation stress ratio (Ishihara et al. 1975), at which the state parameter ψ is also zero. The response of cyclic mobility reflects the occurrence of the phase transformation. Both the loose and dense specimens finally reach a state at which $dp = dq = d\varepsilon_v = 0$ while $d\varepsilon_q \neq 0$. By the classical definition, this is the critical state at which $\psi = 0$.

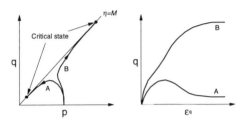

Figure 2.   Influence of density on undrained response of granular materials

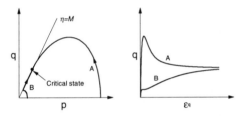

Figure 3.   Influence of initial confining pressure on undrained
response of granular materials

Fig. 3 shows the typical undrained responses of two sand specimens of same density but at different initial effective confining pressures. The specimen subjected to the higher confining pressure is initially in a loose state ($\psi > 0$). The specimen under the lower confining pressure is initially in a dense state ($\psi < 0$). The sand initially in the loose state tends to contract while the sand initially in the dense state tends to dilate. The two specimens finally reach an identical state because of the same void ratio.

Fig. 4 is an illustration of the undrained cyclic response of a medium to dense sand. As the loading cycles proceed, the effective confining stress becomes smaller and smaller, the response also changes from contractive at first few cycles to dilative afterwards. As a comparison, the response to a monotonic stress path is also plotted. It shows that when stress ratio is high, even the confining pressure is not very low, the soil also exhibits phase transformation phenomenon and the corresponding dilative response. It is evident that the dilatancy d, a quantitative measure on whether soil response is contractive (flow liquefaction may occur) or dilative (cyclic mobility may appear), depends on the density as well as the confining pressure and the stress ratio.

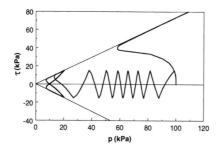

Figure 4. Typical response of medium to dense sand to undrained cyclic simple shear

STATE-DEPENDENT DILATANCY

In the multiaxial stress-strain space, the dilatancy can be defined as follows:

$$d = \frac{d\varepsilon_{kk}^{p}}{\sqrt{\frac{2}{3}de_{ij}^{p}de_{ij}^{p}}} \tag{1}$$

where $e_{ij} = \varepsilon_{ij} - \varepsilon_{kk}\delta_{ij}/3$ are the components of the deviatoric strain tensor. Eq. 1 is consistent with the definition of d in the triaxial compression space. As a function of stress and material state, the dilatancy can be expressed as

$$d = d(R, \theta, \psi, C) \tag{2}$$

where ψ is the state parameter that combines the influence of density and confining pressure and takes critical state as a reference; C represents a set of intrinsic material constants; θ is the Lode angle (Nayak and Zienkiewicz 1972) that defines the direction of shear; and R is a stress ratio invariant defined as

$$R = \frac{\sqrt{3J_{2D}}}{p} = \frac{\sqrt{3s_{ij}s_{ij}/2}}{\sigma_{kk}'/3} \tag{3}$$

where $J_{2D} = s_{ij}s_{ij}/2$ is the second invariant of the deviatoric stress tensor with components $s_{ij} = \sigma_{ij}' - \sigma_{kk}'\delta_{ij}/3$.

In contrast to Eq. 2, the classical stress dilatancy theory in its exact form suggests that the dilatancy d is a unique function of the stress ratio η (the triaxial counterpart of R), i.e.,

$$d = d(\eta, C) \tag{4}$$

While Rowe (1962) clearly pointed out the limitations of this theory and proposed a modification to remove them, the dilatancy is still widely treated by geotechnical engineers as uniquely related to the stress ratio. The modification proposed by Rowe was to include a dependence of d on the internal state of the material. Eq. 2 is exactly such type of modification. It is proven that such a material state dependence is fundamental to the success of unified modeling of the behavior of granular soils.

To see the deficiency of Eq. 4, consider two specimens of the same sand. One is in a loose state and the other in a dense state. Subjected to an incremental shear from the same stress ratio η, the loose specimen contracts and the dense one dilates. These two distinctly different responses are associated with a single η but two different values of dilatancy, one is positive and the other is negative. A constitutive model with this type of dilatancy function cannot properly distinguish the two types of very different responses. As a result, the response of the soil, either contractive or dilative, must be predetermined. In other words, one must predetermine, before a liquefaction analysis, which part of soil may experience flow liquefaction and which part will not.

Furthermore, consider a medium to dense sand subjected to an undrained shear. As shear proceeds, η may pass a phase transformation stress ratio η_{pt} at which $d = 0$ and then approaches the critical state (ultimate state) at which $\eta = M$ and $d = 0$. If Eq. 4 held true, η_{pt} would be equal to a constant M, because d is equal to zero at both the phase transformation state and the critical state. As the critical stress ratio M is an intrinsic material property, there would be a unique phase transformation line for a particular sand, at which the response of the sand changed from contractive to dilative, irrespective to its density and stress level. However, the phase transformation phenomenon can be seen only when the material is in a dense state and in general it is not equal to the critical state stress ratio. As a sand becomes 'looser' due to a lower density or a higher confining pressure, the phase transformation stress ratio η_{pt} becomes higher, and eventually, when the sand becomes too loose, the phase transformation phenomenon totally disappears. The assumption that d is uniquely related to η may lead to erroneous interpretation of soil responses during an earthquake.

Based on the above observations, it is evident that a sand model with its dilatancy formulated within the framework of Eq. 4 will work well only when the variation of the material state is minor. This explains why many sand models have to treat flow liquefaction and cyclic mobility separately, and to treat a sand of different densities as different materials by introducing multiple sets of model parameters.

This dilatancy formulated according to Eq. 2 is termed state-dependent dilatancy for it contains the material state dependence under a fixed η. With this additional dependence, d is now uniquely related to an existing state, a combination of the external stress state and the internal material state. To formulate a particular state-dependent dilatancy, certain rules must be obeyed. Firstly, the dilatancy is restricted by the second law of thermodynamics, and in the absence of residual or back-stress, to obey the following inequality:

$$\delta W^p = p'd\varepsilon_v^p + s_{ij}de_{ij}^p = p'\sqrt{2de_{mn}^p de_{mn}^p/3}\left(R\cos\alpha+d\right)>0 \tag{5}$$

where δW^p is the plastic work increment, and $\alpha=\cos^{-1}\left(s_{ij}de_{ij}^p/|s_{kl}||de_{mn}^p|\right)$ is the angle between s_{ij} and de_{ij}^p. If an associative flow rule is applied in the deviatoric space (Baker and Desai 1982), $\cos\alpha=1$, and Eq. 5 reduced to $R+d>0$. Relation 5 shows that R and d are interrelated at a very fundamental level. Although this requirement is easy to be satisfied, attention needs to be paid when an expression for d is proposed purely based on phenomenological fitting of experimental data.

Secondly, the dilatancy must be zero at critical states, i.e., $d=0$ when $R=M$ and $e=e_c$ (the void ratio at the critical state) or $\psi=0$ simultaneously, i.e.,

$$d(R=M(\theta),\psi=0,C)=0 \tag{6}$$

where $M=M(\theta)$ is the critical state stress ratio in the multiaxial space, which is in general a function of the Lode angle θ.

Thirdly, the state $R=M$ and $\psi=0$ is not the only state at which $d=0$. It is also possible to have a phase transformation state at which $d=0$ but the stress ratio $R_{pt}(\theta)\neq M(\theta)$ and the state parameter $\psi=\psi_{pt}\neq0$, i.e.,

$$d(R=R_{pt}(\theta),\psi=\psi_{pt},C)=0 \tag{7}$$

Eqs. 5, 6 and 7 specify the basic rules to be followed in proposing a particular form of dilatancy function. A particular dilatancy function based on Li and Dafalias (2000) is as follows:

$$d=d_0\left[e^{m\psi}\sqrt{\frac{\overline{\rho}}{\rho}}-\frac{R}{M(\theta)}\right] \tag{8}$$

where d_0, and m are model constants; ρ and $\overline{\rho}$ are projection "distances" in bounding surface formulation. For plastic loading, $\rho=\overline{\rho}$. A simple derivation shows that, when an associative flow rule is used in the deviatoric space, a $d_0<M(\theta)$ guarantees that the relation 5 is satisfied. Observe at critical state, $\psi=0$, $\rho=\overline{\rho}$, and $R=M(\theta)$, one has $d=0$, which satisfies Eq. 6. It can be seen that there are other zero dilatancy states where $\psi\neq0$ and $R\neq M(\theta)$ but $\rho=\overline{\rho}$ and $e^{m\psi}=R/M(\theta)$. Note that for a negative ψ (a dense state), the condition $d=0$ yields a value of $R_{pt}=R_{(d=0,\psi\neq0)}$ smaller than $M(\theta)$. By definition, this R_{pt} is the phase transformation stress ratio. As a function of the state parameter ψ, the phase transformation stress ratio is no longer a constant. This is consistent with Eq. 7.

ELEMENT SIMULATION

A critical state sand model using the above dilatancy function (Eq. 8) has been developed. The model follows the standard bounding surface plasticity theory (Dafalias 1986). A version of the model has successfully simulated data, with a set of unified model parameters (twelve parameters, refer to Table 1), for a suite of seventeen triaxial tests, both drained and undrained, of Toyoura sand over a relative density range from around 0% to 64% subjected to a confining pressure range from 100 kPa to 3000 kPa. Readers are referred to Li and Dafalias (2000) for details.

Similar comparisons between laboratory results and the model simulations on the Leighton Buzzard sand are shown in Figs. 5 to 7. The laboratory tests were carried out recently at the Hong Kong University of Science and Technology, and the model simulations were based on a set of preliminarily calibrated model parameters as listed in Table 1. Fig. 5 shows the undrained response of the sand of various densities under an initial confining pressure of 500 kPa. Fig. 6 shows the undrained response of three loose samples with similar densities but different initial confining pressures. Fig. 7 shows the typical undrained response of the sand of a medium density (Dr=38%) to triaxial cycles. The comparisons demonstrate that the model can effectively capture both the contractive and dilative responses based on the stress and material states.

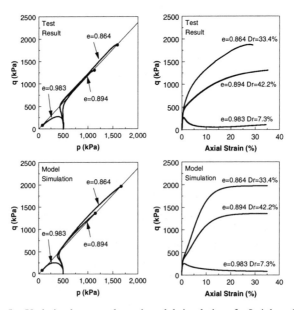

Figure 5.   Undrained test results and model simulations for Leighton Buzzard sand with different densities but the same initial confining pressure

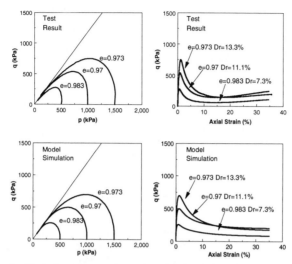

Figure 6. Undrained test results and model simulations for Leighton Buzzard sand with similar densities but different initial confining pressures

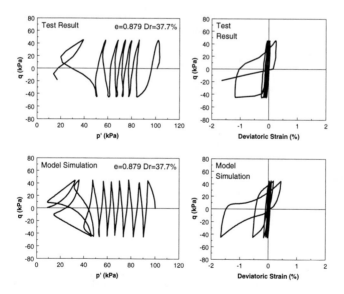

Figure 7. Test results and model simulation for a medium dense Leighton Buzzard sand subjected to undrained cyclic triaxial loading

Table 1. Model parameters for Toyoura sand and Leighton Buzzard Sand

Type of Sand	Elastic Parameters	Critical State Parameters	Dilatancy Parameters	Hardening Parameters
Toyoura sand	$G_0 = 125$ $v = 0.05$	$M = 1.25$ $e_\Gamma = 0.934$ $\lambda_c = 0.019$ $\xi = 0.7$	$d_0 = 0.88$ $m = 3.5$	$h_1 = 3.15$ $h_2 = 3.05$ $h_3 = 3.6$ $n = 1.1$
Leighton Buzzard sand	$G_0 = 150$ $v = 0.05$	$M = 1.185$ $e_\Gamma = 0.998$ $\lambda_c = 0.0198$ $\xi = 0.68$	$d_0 = 0.68$ $m = 1.92$	$h_1 = 2.47$ $h_2 = 2.18$ $h_3 = 7.67$ $n = 1.75$

Flow liquefaction occurs only in loose granular soils with a static driving force. Element simulations were carried out for undrained triaxial compression starting from an anisotropic stress state. Fig. 8 shows the model responses for Toyoura sand of various densities under an initial mean normal effective stress $p_0 = 200$ kPa and a value of $k_0 = 0.6$. It can be seen that at $Dr = 12.4\%$ the sand experiences flow liquefaction with a low residual strength, while at $Dr = 17.6\%$, the sand shows a limited liquefaction response. For the sand of higher relative densities, phase transformation and dilative response are seen clearly.

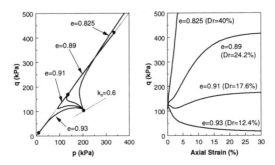

Figure 8.   Model responses to undrained shear for Toyoura sand of different void ratios (initial $k_0 = 0.6$)

Fig. 9 shows the model responses for Toyoura sand at $Dr = 12.4\%$ but with different initial mean normal effective stresses p_0. The initial k_0 value is still fixed at 0.6. It can be seen that flow liquefaction takes place when $p_0 = 50$ kPa and $p_0 = 100$ kPa. In these cases, a strain softening response starts at the peak deviatoric stress and ends with a residual strength on the steady/critical state line. It can also be seen that when p_0 is reduced to 25 kPa, the material is "denser" (the value of state parameter ψ decreases), and a limited liquefaction response shows up.

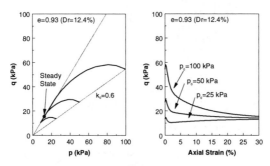

Figure 9.   Model responses to undrained shear for Toyoura sand with the same void ratio but under different initial confining pressures (initial $k_0 = 0.6$)

Figs. 10 and 11 show the model response to undrained cyclic triaxial compression loading for Toyoura sand. Two densities of the sand, Dr=10%, and Dr=37.9%, were simulated. The initial stress state is p_0 = 300 kPa and q = 50 kPa. The cyclic deviatoric stress has a peak-to-peak amplitude of 100 kPa. During loading the lateral total stress is kept constant. Because the loose samples may experience stress softening during loading, a switch was set to change the loading mode from stress control to strain control if stress softening does happen (in that case continuing stress control would become meaningless either physically or numerically). The simulation shows that flow liquefaction has been triggered in the case of Dr=10%. In contrast to this loose sample, the response of the dense sample (Dr=37.9%) shows no stress softening, and as p' decreases, cyclic mobility response shows up. The responses shown in Figs. 10 and 11 demonstrate that the proposed concept can indeed handle both flow liquefaction and cyclic mobility in a unified manner.

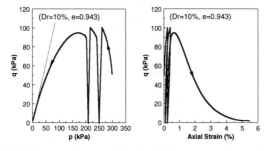

Figure 10. Model response to undrained triaxial compression cycles (Dr=10%)

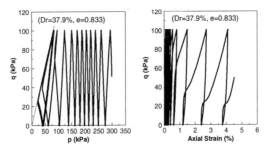

Figure 11.   Model response to undrained triaxial compression cycles (Dr=37.9%)

SEISMIC RESPONSE SIMULATION

The above-described model has been integrated into a fully coupled ground response procedure (Li et al. 1992) to simulate the responses of a hypothetic 10m-thick uniform sand deposit subjected to earthquake input at its bottom. The N-S acceleration time history recorded at 17 m deep in Lotung, Taiwan, during the November 15, 1986, earthquake (M=7.0, refer to Li et al. 1998) was used as the input motion. Level ground and infinitely long sloping ground of 2° and 10° (input motion was applied along the sloping direction) were simulated. The soil deposit was fully saturated. Two densities (Dr=12% and 40%) were used in the simulation. The unified model parameters calibrated for Toyoura sand as listed in Table 1 were used for all the simulations.

Limited by the scope of this paper, only the stress paths and stress-strain responses of a soil element at 5 meters below the ground surface are shown in Figs. 12 to 17. The flow liquefaction and cyclic mobility responses as influenced by the soil density and static driving force can be seen clearly. As shown in Fig. 12, with a low density (Dr=12%) and a large static driving shear stress (larger than the residual strength), flow liquefaction is triggered right after a few cycles of strong shaking. The soil reaches a steady state and enormously large amount of shear strain is developed, causing a significant permanent ground displacement. Fig. 13 shows the response of the same loose soil at the same location but with a smaller static driving shear stress (smaller than the residual strength) attributed to a gentler slope (2°). In this case one sees no flow sliding but only cyclic mobility. The shear deformation is smaller than that caused by flow sliding but is still very significant. Fig. 14 shows the response of the soil element with no static driving force. This type of response (cyclic mobility) is typical for level ground excited by vertically propagating earth shaking. It can be seen that, without a driving force, the shear strain is relatively small and bears only a limited amount of permanent deformation, even though the soil is quite loose. Both Figs 13 and 14 show that as the pore pressure builds up and the mean effective normal stress p' approaches zero, soil gets into a dense state, and correspondingly, the dilative response shows up. Figs. 15 to 17 are the counterparts of Figs. 12 to 14 for the denser soil (Dr=40%) under otherwise identical conditions. As the physical density is higher, the residual strength becomes higher, resulting in that the static driving stresses in all the cases are lower than the material's residual strength. As a result, no flow liquefaction can be seen in any of the cases, and the shear strains are all limited. It is interesting to note that for the sloping ground of 10° (Fig. 15), the stress path reaches the state of phase transformation at a relatively high effective confining pressure p'. This early dilative response slows down the decrease of p' and prevents the fast drop of soil stiffness. Consequently, the shear deformation becomes relatively small (even smaller than its counterpart in the slope of 2°, based on the analysis results). Whereas this mechanism is to be further studied and experimentally verified, the details the presented model can offer are demonstrated.

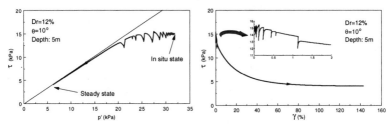

Figure 12.   Response of infinite sloping ground ($\theta = 10°$) at 5m below the surface ($Dr = 12\%$)

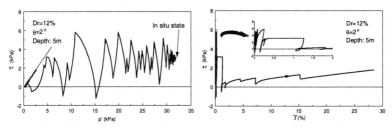

Figure 13.   Response of infinite sloping ground ($\theta = 2°$) at 5m below the surface ($Dr = 12\%$)

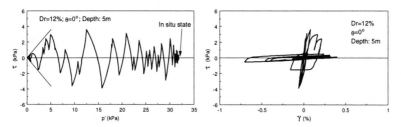

Figure 14.   Response of level ground at 5m below the surface ($Dr = 12\%$)

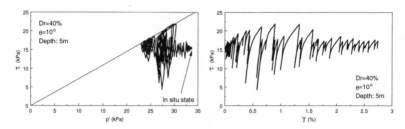

Figure 15.   Response of infinite sloping ground ($\theta = 10°$) at 5m below the surface ($Dr = 40\%$)

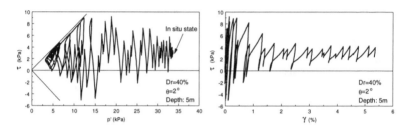

Figure 16.   Response of infinite sloping ground ($\theta = 2°$) at 5m below the surface ($Dr = 40\%$)

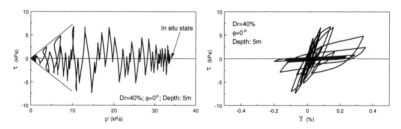

Figure 17.   Response of level ground at 5m below the surface ($Dr = 40\%$)

Although the simulated results are for illustration only, they do show the effectiveness of the proposed modeling framework in conjunction with fully coupled effective stress dynamic procedure.

SUMMARY

The concept of state-dependent dilatancy, which is consistent with the framework of critical state soil mechanics, is described. Theoretical deliberation and numerical simulations show that this simple proposition is effective to meet the challenge in liquefaction analysis. A model formulated within the proposed framework can simulate flow liquefaction as well as cyclic mobility. The stress strain relations either before or after liquefaction depend on the stress and material states. With this framework, there is no need to predetermine which soil is flow liquefiable and which is not, no need to have a rather unnatural liquefaction triggering mechanism, and no need to separately define the behavior of material model for before- and post-liquefaction events.

ACKNOWLEDGEMENT

The financial support provided by the Research Grants Council of Hong Kong through Grant HKUST721/96E is gratefully acknowledged.

APPENDIX. REFERENCES

Baker, R. and Desai, C. S. (1982). "Consequences of deviatoric normality in plasticity with isotropic strain hardening." *Int. J. of Numerical and Analytical Methods in Geomechanics*, 6(3), 383-390.

Been, K. and Jefferies, M. G. (1985). "A state parameter for sands." *Geotechnique*, 35(2), 99-112.

Castro, G. (1969). "Liquefaction of sands." *Harvard Soil Mechanics Series 87*, Harvard University, Cambridge, Massachusetts.

Chan, D. H. and Morgenstern, N. R. (1989). "Bearing capacity of strain-softening soil." De Mello Vol., Editora Edgard Blucher Ltda., Sao Paulo, Brazil, 59-68.

Cubrinovski, M. and Ishihara, K. (1998). "Modeling of sand behavior based on state concept." *Soils and Foundations*, 38(3), 115-127.

Dafalias, Y. F. (1986). "An anisotropic critical state soil plasticity model." *Mechanics Research Communications*, 13(6), 341-341.

Finn, W. D. Liam. (1990). "Analysis of post-liquefaction deformations in soil structures." *Proceedings of H. Bolton Seed Memorial Symposium*, Duncan J. M. (editor), University of California, Bi-Tech Publishers, Vancouver, Canada, 2, 291-311.

Gu, W. H., Morgenstern, N. R., and Robertson, P. K. (1993). "Postearthquake deformation analysis of Wildlife site." *Journal of Geotechnical Engineering*, ASCE, 120(2), 274-289.

Gu, W. H., Morgenstern, N. R., and Robertson, P. K. (1994). "Progressive failure of lower San Fernando dam." *Journal of Geotechnical Engineering*, ASCE, 119(2), 333-348.

Ishihara, K., Tatsuoka, F, and Yasuda, S. (1975). "Undrained deformation and liquefaction of sand under cyclic stresses." *Soils and Foundations*, 15(1), 19-44.

Li, X. S., Wang, Z. L., and Shen, C. K. (1992). "SUMDES, a nonlinear procedure for response analysis of horizontally-layered sites subjected to multi-directional earthquake loading." *Report to the Department of Civil Engineering*, University of California, Davis.

Li, X. S. (1997). "Modeling of dilative shear failure." *Journal of Geotechnical Engineering*, ASCE, 123(7), 609-616.

Li, X. S., Shen, C. K., and Wang, Z. L. (1998). "Fully-coupled inelastic site response analysis for 1986 Lotung earthquakes." *Journal of Geotechnical and Geoenvironmental Engineering*, ASCE, 124(7), 560-573.

Li, X. S., Dafalias, Y. F., and Wang, Z. L. (1999). "State dependent dilatancy in critical state constitutive modeling of sand." *Canadian Geotechnical Journal*, 36(4), 599-611.

Li, X. S. and Dafalias, Y. F. (2000). "Dilatancy for cohesionless soils." *Geotechnique*, in print.

Manzari, M. T. and Dafalias, Y. F. (1997). "A critical state two-surface plasticity model for sands." *Geotechnique*, 47(2), 255-272.

Nayak, G. C., and Zienkiewicz, O. C. (1972). "A convenient form of invariants and its application in plasticity." *J. of Soil Mech. and Foundations*, ASCE, 98(4), 949-954.

Rowe, P. W. (1962). "The stress-dilatancy relation for static equilibrium of an assembly of particles in contact." *Proc. Royal Society*, London, A269, 500-527.

Scott, R. F. and Arulanandan, K. (1993). *Verification of Numerical Procedures for the Analysis of Soil Liquefaction Problems*, 1 & 2, A. A. Balkema Publishers, Rotterdam.

Sladen, J. A., D'Hollander, R. D., and Krahn, J. (1985). "The liquefaction of sand, a collapse surface approach." *Canadian Geotechnical Journal*, 22(4), 564-578.

Verdugo, R. and Ishihara, K. (1996). "The steady state of sandy soils." *Soils and Foundations*, 36(2), 81-91.

Wan, R. G. and Guo, R. G. (1998). "A simple constitutive model for granular soils: modified stress-dilatancy approach." *Computers and Geotechnics*, 22(2), 109-133.

Wang, Z. L., Dafalias, Y. F., and Shen, C. K. (1990). "Bounding surface hypoplasticity model for sand." *Journal of Engineering Mechanics*, ASCE, 116(5), 983-1001.

DSC Constitutive and Computer Models
for Soil-Structure and Liquefaction Analysis

Chandrakant Desai[1], F. ASCE

Abstract

This paper presents a unified and simplified constitutive modeling approach, called the disturbed state concept (DSC), for the characterization of the mechanical behavior of soils and interfaces. It is calibrated and validated with respect to laboratory stress-strain-volume change-pore water pressure behavior. Finite element procedures based on generalized Biot's theory for coupled response of saturated materials is developed with the DSC model. It is applied to a number of soil-structure interaction problems in which the predictions in terms of displacements, stresses and pore water pressures, and liquefaction potential are compared with laboratory or field observations. It is believed that the DSC can provide improved constitutive models towards analysis and design of geotechnical problems, and a fundamental procedure for identification of liquefaction potential.

Introduction

Soil-structure interaction has significant effect on the behavior of many geotechnical problems subjected to static and dynamic (earthquake) loads. Nonlinear response of soils and interfaces, initial conditions, type of loading, plastic hardening, relative particle motions or microcracking leading to degradation or softening, and stiffening or hardening, are among the factors that can influence the soil-structure response. Although various stress-strain or constitutive models have been proposed for soils, there appears to be no models available that can account for the above factors in a unified manner. The behavior at the interfaces between structure and geologic materials involves different deformation mechanism than that in the soil (solid) material. Here again, a constitutive

---

[1] Regents' Professor, Dept. of Civil Eng. and Eng. Mech., The Univ. of Arizona, Tucson, AZ 85721

modeling framework for interfaces that is consistent with that for the soil, has not been developed.

Under loading, a saturated sand with certain range of (initial) density may experience microstructural particle motions that can lead to states of local and global instabilities. The latter can lead to liquefaction and resulting failure of the foundation soil. A great number of empirical and mechanistic models have been proposed and developed for liquefaction in sands. However, the issue of liquefaction at and in the vicinity of the interface, that can occur before or after that in the neighboring soils, has not been investigated before.

Scope

The scope of this paper entails the following items:

(a) A unified constitutive model, called the disturbed state concept (DSC), that allows for elastic, plastic and creep responses, relative particle motions and microcracking leading to softening and liquefaction, and stiffening or healing,

(b) Use of the DSC model for soils and interfaces,

(c) Laboratory testing for calibration of parameters for soils and interfaces,

(d) Validation of DSC models for soils and interfaces,

(e) Implementation of DSC model is nonlinear coupled static and dynamic finite element procedure, and

(f) Verification and analysis of typical problems tested in the laboratory and field.

In view of the length and time limitations, the descriptions presented below are brief, and related directly to the DSC. As a result, no detailed review of other literature is included; it is available in various references cited.

The Disturbed State Concept

The DSC represents a unified constitutive modeling approach that allows characterization of the behavior of "solids" (soils, rocks, concrete, ceramics, metal alloys), and interfaces and joints with the same mathematical framework as for solids. It permits, in a hierarchical manner, elastic, plastic and creep responses, relative particle motions, microcracking and fracture leading to softening, and healing or stiffening under thermomechanical loading.

The DSC possesses a number of advantages compared to other available models:

(1) Its hierarchical character provides the user the flexibility to choose a version(s), elastic, elastoplastic, viscoplastic and disturbance (damage and

softening), depending on the behavior of a given material(s) for specific application. In other words, only parameters relevant to the selected version are required to be input.

(2) It is based on fundamental mechanistic considerations, and yet is simplified for practical applications. For example, for a given capability, it involves lesser number of parameters compared to other available models.

(3) Its parameters have physical meaning and can be determined from standard laboratory tests.

Details of the DSC and its hierarchical (hierarchical single surface – HISS plasticity) versions are given elsewhere (Desai, 1995, 1999; Desai and Ma, 1992; Desai, et al., 1997a, 1998a, 1998b; Desai and Toth, 1996; Katti and Desai, 1995; Shao and Desai, 1998; Park and Desai, 1999). The basic constitutive equations for the DSC are given below.

$$d\sigma^a = (1-D)d\sigma^i + D\sigma^c + dD(\sigma^c - \sigma^i) \tag{1a}$$

or
$$d\sigma^a = (1-D)C^i d\varepsilon^i + DC^c d\varepsilon^c + dD(\sigma^c - \sigma^i) \tag{1b}$$

where σ and ε = stress and strain vectors, respectively, a, i and c denote observed, RI and FA states, respectively, d denotes increment or rate, C = constitutive matrix and D = (scalar) disturbance, which can be expressed as a tensor if appropriate laboratory tests are available (Desai and Toth, 1996). In the DSC, the stresses and strains in the RI and FA states can be different; however, if they are assumed to be compatible (i.e., $d\varepsilon^a = d\varepsilon^i = d\varepsilon^c$), the formulation is simplified and only Eqs. (1) are required. Equations (1) include elastic, elastoplastic (or viscoplastic) models or versions as special cases when D = 0. If D ≠ 0, disturbance (microcracking and damage) is allowed, Desai and Toth (1996), Desai (1995, 1999).

In the DSC, it is assumed that during deformation, the material transforms from the initial *continuum* or relative intact (RI) state to the fully adjusted (FA) state as a result of continuing changes in the material's microstructure. Hence, at any state during deformation, the material element is composed of a mixture of parts in the RI and FA states. The observed or actual response of the material is then expressed in terms of the behavior under the RI and FA states through disturbance (D), which acts as an interpolation and coupling mechanism (Desai, 1995, 1999).

The RI response can be characterized by using such continuum theories as elasticity, elastoplasticity and viscoplasticity, with thermal effects. Here, the elastoplastic model in the HISS approach (Desai, et al., 1986) is used. In the FA state, the material parts can carry no stress at all, as they act as cracks or voids, like in the classical damage model, or it can carry hydrostatic stress but no shear stress like a constrained liquid, or it can carry shear stress reached up to that state for a given mean pressure and deform in shear without volume change as in the critical state concept (Roscoe, et al., 1957; Desai, 1995, 1999). The latter two are considered to be realistic and provide for important interaction between the RI and FA parts, and are used often. In this paper, the FA state is characterized by using the critical state concept.

The DSC formulation includes in its framework the coupling between the RI and FA parts; as a result, it is not necessary to add external enrichments such as microcrack interaction, Cosserat and gradient theories. Because of this, the DSC model allows for nonlocal effects and avoids spurious mesh dependence (Desai, et al., 1997b; Desai, 1999).

Unloading and Reloading

Details of the simplified procedures to simulate unloading and reloading are given elsewhere (Desai, et al., 1997a; Shao and Desai, 1998). Their characterization requires two additional parameters, the slope of stress-strain curve at the end of unloading (E^u) and the irreversible strains during unloading cycle (ε^p).

Interfaces

The foregoing formulation for solids can be specialized to characterize behavior of interfaces (or joints) idealized as thin layers between structural and geologic materials (Desai, et al., 1984). The thin-layer interface element is treated as a "solid" element; however, the parameters for the DSC model are determined on the basis of static or cyclic interface tests. Details are given by Desai and Ma (1996), Desai and Fishman (1991), Desai and Rigby (1997), Park and Desai (1999), and Shao and Desai (1998).

Parameters

The parameters in the DSC model have physical meanings as they are related to specific states during deformation. Their number is lower than that in other available models of comparable capabilities. For example, for the general disturbance model that allows for elastic and plastic strains including continuous yielding, and microcracking leading to softening, the parameters involved are two elastic (E, v); four plasticity (γ, β, a_1 and η_1), three critical state (e_o, λ, $\overline{m}$) and three for disturbance (A, Z and D_u). These parameters can be found from standard triaxial (or multiaxial) tests; the disturbance parameters are found based on the softening or cyclic degradation response.

Validations

The DSC model and its versions have been applied successfully to characterize behavior of cohesionless soils, cohesive saturated soils, rocks, concrete, ceramic composites, metal alloys like solders in electronic packaging problems, and silicon with dislocation and impurities, and interfaces and joints (Desai, 1995, 1998; Desai and Fishman, 1991; Desai and Ma, 1992; Katti and Desai, 1995; Desai and Toth, 1996; Desai, et al., 1998a, 1998b; Desai and Rigby, 1997; Park and Desai, 1997; Shao and Desai, 1998). The model has been implemented in static and dynamic finite element procedures for dry and saturated materials and interfaces (Desai, et al. 1997; Shao and Desai, 1998; Park and Desai, 1999).

Finite Element Procedure

The generalized Biot's theory is used to formulate the finite element (FE) procedure for coupled deformation and pore water pressure responses. The procedure allows different versions; e.g., dynamic, consolidation and static. The DSC model for loading, unloading and reloading is implemented in the FE procedure. The results from the code include time dependent displacements, strains, stresses (total and effective), pore water pressures and disturbance. Contours of disturbances with time in the finite element mesh provide identification of zones in which the critical disturbance, D_c, is reached (Desai, et al., 1998b; Desai, 2000). It denotes initiation of liquefaction, and allows tracing of the growth of disturbance and liquefaction under subsequent loading cycles.

Applications

A number of problems involving static, repetitive and dynamic loading with elastoplastic, viscoplastic and DSC models have been solved by using the computer code. In most cases, the computer predictions are compared with observations in the field and simulated laboratory models. Here, brief descriptions of three typical examples involving a pile in clays, pile in sand, and dynamic soil-structure interaction in a shake table test are presented.

Pile in Marine Clay

Instrumented pile segments were tested in the field at Sabine, Texas by Earth Technology Corporation (ETC, 1986). The field program included tests for measurement of *in situ* stresses, installation of pile segments with different diameters and cutting shoes, monitoring consolidation, performing axial tension tests at different levels of consolidation and cyclic axial load tests at the end of consolidation. The measurements included total lateral stresses and pore water pressures at the pile wall, and shear transfer versus pile displacement. Undisturbed specimens of soil were obtained by using Shelby tubes and specially designed rectangular tubes for cyclic triaxial and multiaxial testing of the marine clay, respectively. Interface tests between the marine clay and pile (steel) were conducted by using the cyclic multi degree-of-freedom shear device (Desai and Rigby, 1997).

The results presented here are based on the coupled finite element procedure including the disturbed state model for loading, unloading and reloading (Desai, et al., 1997; Shao and Desai, 1998). Various sequences such as initial conditions, pile driving, consolidation, axial tension tests and cyclic loading were simulated in the finite element analysis. The FE mesh is shown in Fig. 1. The inside nodes on the soil-pile interface (along EF) were subjected to cyclic (vertical) displacements as it was applied in the field.

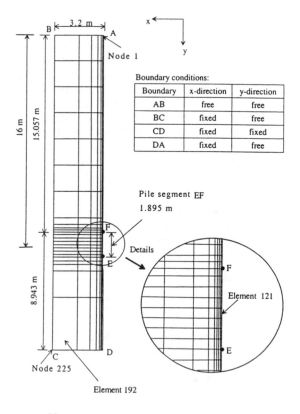

Figure 1. Finite Element Mesh of Field Pile Test

Figures 2 (a) to (c) show typical comparisons between the predictions in terms of shear transfer versus displacement, and shear transfer and pore water pressures versus time from the DSC model and field behavior under two-way cyclic loading, respectively. They also show predictions by using anisotropic hardening plasticity (HISS) model without the provision for degradation (disturbance) (Wathugala and Desai, 1993; Desai, et al., 1993). It can be seen that the DSC predictions correlate very well with the field data, and the provision for degradation (disturbance) provides improved predictions compared to those from the HISS-plasticity model.

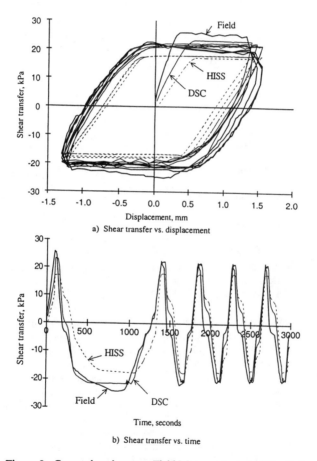

a) Shear transfer vs. displacement

b) Shear transfer vs. time

Figure 2. Comparison between Field Measurements and Predictions From DSC and HISS Models: Two-way Cyclic Load Test

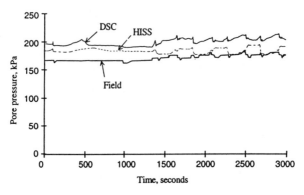

c) Pore pressure vs. displacement

Figure 2 (continued)

Pile in Saturated Sand

In order to study the interaction and liquefaction behavior, a steel pile in saturated Ottawa sand was simulated by using the DSC model and the finite element procedure. Figure 3(a) shows the pile-sand details, and Fig. 3(b) shows mesh details near the interface in which sinusoidal displacement loading was applied at the nodes (Park and Desai, 1997).

The cyclic behavior of the soil was characterized from a series of tests using the multiaxial test device with 10 x 10 x 10 cm saturated specimens (Gyi, 1996). The interface DSC behavior was characterized based on a series of interface tests between steel and sand using the CYMDOF-P device (Desai and Rigby, 1997).

Computer analyses were performed for two conditions: (1) no interface, i.e., pile and soil are compatible, and (2) with interface, i.e., relative motions are allowed between the soil and pile. Figures 4(a) and (b) show computed (vertical) displacements with time at typical notes 136 and 137; the former is on the pile and the latter in the soil. Figures 5(a) and (b) show computed pore water pressures in typical (soil) element 121 with time. Figures 6(a) and (b) show disturbance in soil elements 121 and 122, Fig. 3(b), with time. It can be seen that provision of the interface, i.e., relative motions, modify the computed results significantly. There occur significant relative displacements between pile and soil with the interface provision. The variation in the magnitudes of cyclic pore water pressures are much lower with the interface, indicating that the relative motions cause "damping" in the pore pressures generated. The disturbances in the vicinity of the interface are generally smaller with the interface.

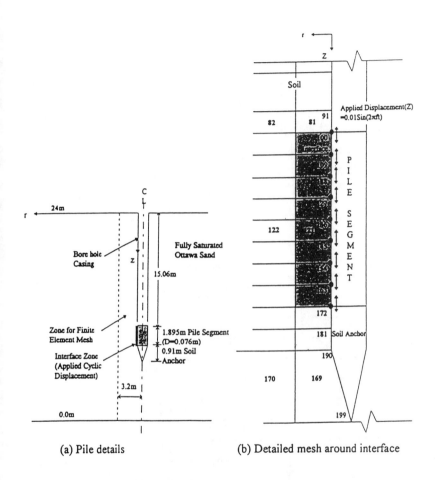

(a) Pile details

(b) Detailed mesh around interface

Figure 3. Pile in Saturated Sand (Park and Desai, 1997)

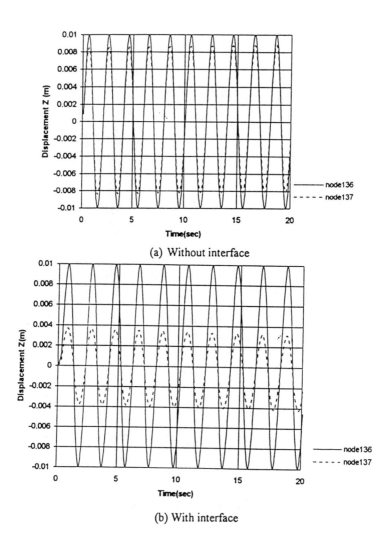

(a) Without interface

(b) With interface

Figure 4. Displacement at Typical Adjacent Nodes near Interface

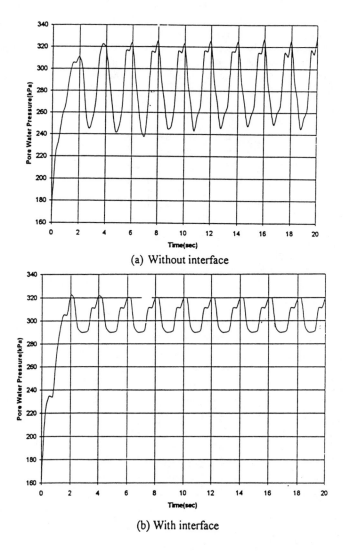

(a) Without interface

(b) With interface

Figure 5. Pore Water Pressure in Element 121

Based on laboratory tests (Gyi, 1996; Park and Desai, 1999; Desai, et al., 1998b), it was found that liquefaction initiated at the average critical disturbance of about 0.84. Results without interface, Fig. 6(a), shows that the value of D_c is reached after about two seconds (i.e., four cycles of loading, frequency = 0.5 Hz) in both elements. On the other hand, results with interface, Fig. 6(b), shows that the critical disturbance (at liquefaction), D_c ($\approx$ 0.84), is reached in the vicinity of the interface after about four cycles, whereas that in the nearby soil element liquefaction is indicated after about twenty-four cycles. Thus, the provision interface can allow identification of liquefaction in the interface zone.

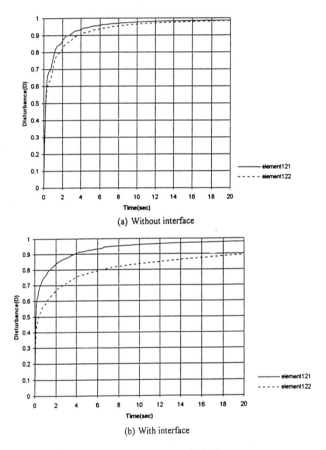

(a) Without interface

(b) With interface

Figure 6. Disturbances in Typical Elements

Shake Table Test: Soil-Structure Interaction

Akiyoshi, et al. (1996) reported shake table tests with a saturated sand. Figure 7(a) shows the details of the model, which was subjected to cyclic sinusoidal displacement (x) at the bottom of the test box, given by

$$X = \bar{x} \sin (2\pi \, ft) \qquad (2)$$

where $\bar{x}$ = 0.0013 m, frequency f = 5.0 HZ and t = time.

The finite element mesh is shown in Fig. 7(b); the idea of repeating side boundaries was used. The material parameters for the Ottawa sand used were obtained from multiaxial cyclic tests (Gyi, 1996; Park and Desai, 1999).

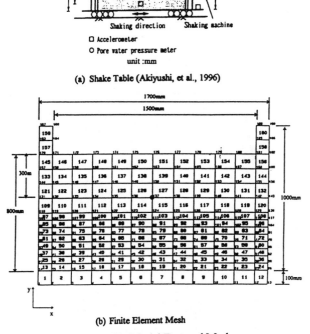

(a) Shake Table (Akiyushi, et al., 1996)

(b) Finite Element Mesh

Figure 7. Set Up Model Test and Mesh

Figure 8 shows comparisons between computed and observed pore water pressures at node (137) at the depth of 300 mm. Figure 9 shows the growth of disturbance near node 137 with time, and indicates the critical disturbance, D_c = 0.84, at which liquefaction initiated (Desai, et al., 1998b). This compares well with the observations in which it was found that liquefaction occurred at time ≈ 2.00 secs, Fig. 8. Plot of contours of disturbance at different time (t = 0.50, 1.0, 2.0, 10.0 secs) showed that liquefaction initiated in the zones at and below the depth = 300 mm when D_c ≈ 0.84 was reached (Park and Desai, 1999). Then, as the disturbance increased with cycles beyond D_c = 0.84, liquefaction expanded essentially in the entire zone at t = 10 secs, as was observed in the laboratory test.

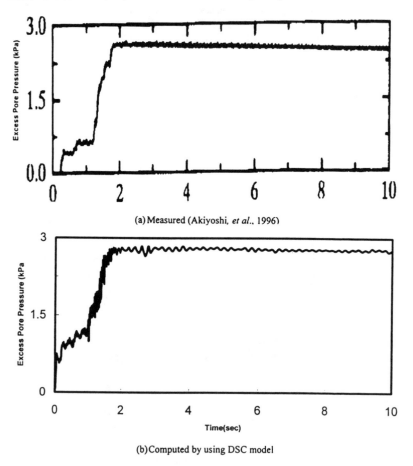

(a) Measured (Akiyoshi, *et al.,* 1996)

(b) Computed by using DSC model

Figure 8. Excess Pore Pressure at Depth of 300 mm

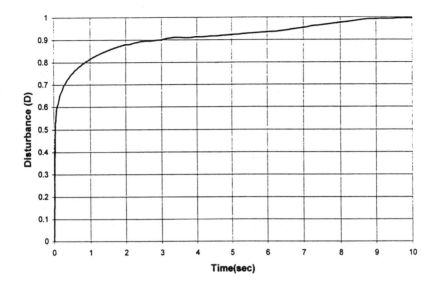

Figure 9. Growth of Disturbance in Sand and at Depth = 300 mm

Summary and Conclusions

The recently developed unified and hierarchical model based on the disturbed state concept (DSC) is described. The required material parameters, their determination from laboratory tests and validations for a wide range of materials and interfaces, are discussed. The DSC model is implemented in a general nonlinear finite element procedure. Typical practical problems involving piles in clays and sands and shake table test are solved by using the numerical procedure. The numerical predictions are compared with observed data in the field and laboratory tests. The subject of liquefaction during dynamic loading is considered and the idea of the critical disturbance in the DSC is used to identify initiation of liquefaction and its growth. This approach is considered to be fundamental compared to other empirical approaches for liquefaction potential (Desai, et al., 1998b; Desai, 2000).

It is believed that the proposed unified DSC constitutive model for soils and interfaces and the associated computer procedure can provide improved analysis and design of dynamic and static soil-structure problems in geotechnical engineering.

Acknowledgments

Parts of the research results were obtained under various grants (Nos. CSM 9115316, CES 8711764, MSM 8618914) from the National Science Foundation, Washington, DC. A number of students and coworkers contributed (see references cited) to the results presented herein; assistance of Drs. C. Shao and I.J. Park is acknowledged.

References

Akiyoshi, T., Fang, H.L., Fuchida, K., and Matsumoto, H. (1996). "A nonlinear seismic response analysis method for saturated soil-structure system with absorbing boundary." *Int. J. Num. and Analyt. Meth. Geomech.*, 20(5), 307-329.

Desai, C.S. (1995). "Constitutive modelling using the disturbed state as microstructure self-adjustment concept." Chapter 8 in *Continuum models for materials with microstructure*. H.B. Mühlhaus (editor), John Wiley, Chichester, United Kingdom.

Desai, C.S. (1999). *Mechanics of materials and interfaces: the disturbed state concept.* CRC Press, Boca Raton, under publication.

Desai, C.S. (2000). "Evaluation of liquefaction using disturbed state and energy approaches." *J. of Geotech. and Geoenv. Eng.*, ASCE, in press.

Desai, C.S., Basaran, C., Dishongh, T., and Prince, J. (1998a). "Thermomechanical analysis in electronic packaging with unified constitutive model for materials and joints." *Components, Packaging and Manuf. Tech., Part B: Advanced Packaging*, IEEE Trans., 21(1), 87-97.

Desai, C.S., and Fishman, K.L. (1991). "Plasticity based constitutive model with associated testing for joints." *Int. J. Rock Mech. and Min. Sc.*, 28(1), 15-26.

Desai, C.S., and Ma, Y. (1992). "Modelling of joints and interfaces using disturbed state concept." *Int. J. Num. Analyt. Meth. Geomech.*, 16(9), 623-653.

Desai, C.S., Park, I.J., and Shao, C. (1998b). "Fundamental yet simplified model for liquefaction instability." *Int. J. Num. Analy. Meth. Geomech.*, 22, 721-748.

Desai, C.S., and Rigby, D.B. (1997). "Cyclic interface and joint shear device including pore pressure effects." *J. of Geotech. And Geoenv. Eng.*, ASCE, 123(6), 568-579.

Desai, C.S., Shao, C., and Park, I.J. (1997a). "Disturbed state modelling of cyclic behavior of soils and interfaces in dynamic soil-structure interaction." *Proc., 9th Int. Conf. on Computer Meth. and Advances in Geomech.*, Wuhan, China.

Desai, C.S., Basaran, C., and Zhang, W. (1997b). "Numerical algorithms and mesh dependence in the disturbed state concept." *Int. J. Num. Meth. Eng.*, 40, 3059-3083.

Desai, C.S., Somasundaram, S., and Frantziskonis, G. (1986). "A hierarchical approach for constitutive modeling of geologic materials." *Int. J. Num. Analyt. Meth. Geomech.*, 10(3), 225-257.

Desai, C.S., and Toth, J. (1996). "Disturbed state constitutive modelling based on stress-strain and nondestructive behavior." *Int. J. Solids and Struct.*, 33(11), 1619-1650.

Desai, C.S., Wathugala, G.W., and Matlock, H. (1993). "Constitutive model for cyclic behavior of cohesive soils II: applications." *J. Geotech. Eng.*, ASCE, 119(4), 730-748.

Desai, C.S., Zaman, M.M., Lightner, J.G., and Siriwardane, H.J. (1984). "Thin-layer element for interfaces and joints." *Int. J. Num. Analyt. Meth. Geomech.*, 8, 19-43.

Earth Technology Corporation (1986). "Pile segment tests – Sabine Pass: some aspects of the fundamental behavior of axially loaded piles in clay soils." *ETC Report No. 85-007*, Houston, Texas.

Gyi, M.M. (1996). "Multiaxial cyclic testing of saturated Ottawa sand," *M.S. Thesis*, Dept. of Civil Eng. and Eng. Mechs., The Univ. of Arizona, Tucson, Arizona.

Katti, D.R., and Desai, C.S. (1995). "Modelling and testing of cohesive soil using disturbed state concept." *J. of Eng. Mech.*, ASCE, 121(5), 648-658.

Park, I.J., and Desai, C.S. (1997). "Disturbed state modeling for dynamic and liquefaction analysis." *Report to NSF*, Dept. of Civil Eng. and Eng. Mech., The Univ. of Arizona, Tucson, Arizona.

Park, I.J., and Desai, C.S. (1999). "Cyclic behavior and liquefaction of sand using disturbed state concept." *J. of Geotech. and Geoenv. Eng.*, ASCE, in press.

Roscoe, K.H., Schofield, A., and Wroth, C.P. (1958). "On the yielding of soils." *Geotechnique*, 8, 22-53.

Shao, C., and Desai, C.S. (1998). "Implementation of DSC model for dynamic analysis of soil-structure interaction problems." *Report to NSF*, Dept. of Civil Eng. and Eng. Mechs., The Univ. of Arizona, Tucson, Arizona.

Wathugala, G.W., and Desai. C.S. (1993). "Constitutive model for cyclic behavior of cohesive soils I: theory." *J. Geotech. Eng.*, ASCE, 119(4), 714-729.

Fully-Coupled Analysis of a Single Pile Founded in Liquefiable Sands.

A. Anandarajah [1]

Abstract

Using a fully-coupled finite element analysis method, the earthquake be-havior of a single pile subjected to an earthquake base shaking in a centrifuge is analysed. An effective stress based elasto-plastic constitutive model is used to represent the stress-strain behavior of the sand. A few years ago, this model was used in a before-the-event prediction exercise and the numerical predictions were found to be reasonably close to observations; samples of com-parison between predictions and data are presented. The pile-soil interaction problem is analysed using a two-dimensional model of the problem, and it is shown that the results from the fully-coupled procedure compare reasonably well with centrifuge data.

Introduction

Prediction of the behavior of piles and pile groups during earthquakes still remains a difficult task, especially when the soil supporting the piles expe-riences liquefaction. Two broad classes of approaches are currently being pursued by researchers: (a) Nonlinear Winkler Foundation Method that uses nonlinear p-y stiffness to represent the soil, and (b) finite element numerical method. This paper concerns the latter. Within the general framework of the finite element methods, several variations are again possible: total stress approaches based on equivalent linear soil properties, loosely-coupled effective stress approaches, fully-coupled approaches, etc. The analyses presented here have been performed using a fully-coupled method. The method allows the

---

[1]Professor, Department of Civil Engineering, The Johns Hopkins University, Baltimore, Maryland, USA

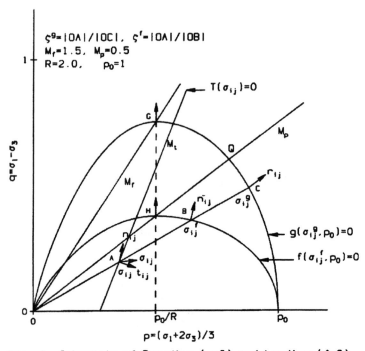

Fig. 1. Schematic of Bounding (g=0) and Loading (f=0) Surfaces, and Conical Surface (T=0).

Figure 1: Schematic of the Two Surfaces Used for Loading and the Conical Surface Used for Stress Reversal

deformation and pore water pressure variations to be realistically simulated by modeling the pore pressure build-up and dissipation simultaneously. While 3D analyses by this approach are still CPU intensive, 2D analyses may now be performed within reasonable time periods. In view of the rate at which the computers are becoming faster and faster, even 3D analyses may be routinely performed in the near future. The analyses presented here have been performed on two-dimensional idealized finite element models and a series of three-dimensional analyses are currently being performed.

Constitutive Model

A two-surface plasticity model was recently developed for describing the stress-strain relationships of sands under both monotonic and cyclic loading;

the latter under liquefying and non-liquefying conditions (Anandarajah, 1994a; 1994b); the reader is referred to these references for full details. Here, only the salient features of the model are pointed out.

The two surfaces employed in the model are shown in a triaxial space in Fig. 1; an outer surface $g = 0$ referred to as the bounding surface and an inner surface $f = 0$ referred to as the loading surface. An ellipse and distorted ellipse are used to represent $g = 0$ for $M_f \geq \frac{q^g}{p^g} \geq 0$ and $\frac{q^g}{p^g} \geq M_f$ respectively. The slope M_f is assumed to be a constant, whereas M_p is assumed to vary during loading, becoming equal to M_f when the stress state reaches $g = 0$. $f = 0$ and $g = 0$ coincide at this stage. The variable p_0, which defines the size of bounding surface (Fig. 1), is assumed to change as a function of the plastic internal variables ϵ_v^p and ξ^p as described subsequently.

The plastic strain rate is expressed as

$$\dot{\epsilon}_{ij}^p = L b_{ij} \tag{1}$$

where

$$L = \frac{1}{K_p} \dot{\sigma}_{k\ell} b_{k\ell} \tag{2}$$

K_p is the plastic modulus, and $(b_{k\ell} b_{k\ell})^{\frac{1}{2}} = 1$. The index L is used to separate "loading" from "stress reversal". It is noted that the model uses associated flow rule. A stress probing is defined as **loading** when $L > 0$, **neutral loading** when $L = 0$ and **stress reversal** when $L < 0$. It is understood that when $K_p > 0$, a stress probing will either be loading or stress reversal depending on whether $\dot{\sigma}_{k\ell} b_{k\ell} > 0$ (i.e., the direction of $\dot{\sigma}_{k\ell}$ is directed outward from the surface that $b_{k\ell}$ is normal to) or $\dot{\sigma}_{k\ell} b_{k\ell} < 0$ (i.e., the direction of $\dot{\sigma}_{k\ell}$ is directed inward from the surface that $b_{k\ell}$ is normal to). However, when $K_p < 0$, $\dot{\sigma}_{k\ell} b_{k\ell} < 0$ is still defined as loading. This will be the case during strain-softening of dense sands and static liquefaction of very loose sands (Castro, 1969).

The behavior is always modelled to be **elasto-plastic** when $L > 0$ and **elastic** when $L \leq 0$.

The inner surface $f = 0$ is used to define the direction of plastic strain. First, L is evaluated using K_p and n_{ij} (Fig. 1). When this L is positive, the event is assumed to be loading, and b_{ij} is set equal to n_{ij}. When L is negative, the event is taken to be stress reversal, and K_p and b_{ij} are re-defined. Defining a cone $T = 0$ with I−axis taken as its axis, $b_{ij} = t_{ij}$ is taken to be the inward normal to $T = 0$, as shown in Fig. 1.

Two plastic internal variables, ϵ_v^p and ξ^p are employed in defining the hardening/softening behavior of the material. ϵ_v^p and ξ^p changes during loading, but are assumed to remain constant during stress reversal regardless of whether a stress probe during stress reversal causes elastic or elasto-plastic behavior. It may also be noted that ξ^p either increases or remains constant (since $\dot{\xi} \geq 0$), whereas ϵ_v^p either increases or decreases depending on whether $\dot{\epsilon}_v^p > 0$ (compaction) or $\dot{\epsilon}_v^p < 0$ (dilation).

While the surface $f = 0$ is used to define loading/stress reversal events, the plastic modulus is defined with respect to the surface $g = 0$. Let K_b be the plastic modulus on the surface at any point such as point C. K_b is referred to as the bounding modulus (Dafalias and Herrmann, 1986). The plastic modulus at a point within $g = 0$ (e.g., point A) is related to K_b as

$$K_p = K_b + H(\zeta^g, \sigma_{ij}) \tag{3}$$

where ζ^g is a scalar distance, defined in Fig. 1, whose value varies between 0 at point O (origin) to 1 on $g = 0$. The modulus H is referred to as the shape-hardening modulus (Dafalias and Herrmann, 1986), which is defined so that $H(\zeta^g = 1, \sigma_{ij}) = 0$. Having selected a hardening rule, K_b is determined by the consistency condition applied to the bounding surface $g = 0$.

The evolution of p_0 is assumed to depend on both the volumetric and deviatoric plastic strains as

$$\dot{p}_0 = \frac{1 + e_0}{\lambda^*} p_0 \dot{\epsilon}_v^p + \beta(\xi^p, p_0) p_0 \dot{\xi}^p \tag{4}$$

where,

$$\beta = (\gamma_0 - m \frac{p_0}{p_a}) e^{-\gamma_1 \xi^p} \tag{5}$$

and $\lambda^* = \lambda - \kappa$, λ and κ are slopes of isotropic compression and rebound lines in the $e - \ln p$ space. e is the void ratio and e_0 is its initial value. γ_0, m and γ_1 are model parameters, defined in general to be positive. γ_0 and m are to be selected such that β will also be positive. β decreases exponentially with increasing ξ^p and linearly with increasing p_0. ξ^p and p_0 do not change during stress reversal. As a result, β remains a constant during stress reversal, but begins to change again when a stress probe causes loading.

For a state of stress lying within $g = 0$, $M_p \neq M_f$ in general and thus $g \neq f$. M_p is assumed to vary from $\alpha_g M_f$ to M_f ($0 \leq \alpha_g \leq 1$) as ζ^g varies from 0 to 1 as

$$M_p = M_f[\alpha_g + (1 - \alpha_g)(\zeta^g)^{n_g}] \tag{6}$$

where α_g and n_g are material parameters.

The equations used to model the behavior upon stress reversal are summarized below:

$$K_p^* = -h_r K_{b0} \tag{7}$$

$$K > K_{min} \tag{8}$$

$$K_{min} = \frac{1.1 K_{p0} M_{t0}}{(M_{t0} - M_p) t_{p0}^2} \tag{9}$$

$$\frac{1}{M_{t0}} = \frac{0.9}{M_p}\left[1 - e^{-\alpha_r \frac{\eta^{max}}{<1 - \eta^{max}>}}\right] \tag{10}$$

$$M_t = \frac{M_{t0}}{\eta^f} \tag{11}$$

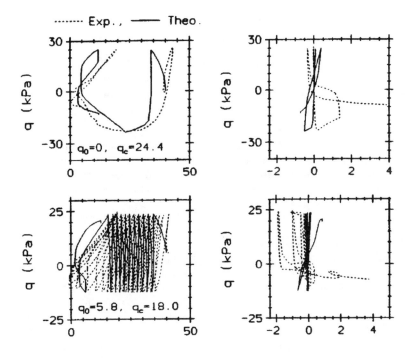

Figure 2: Comparison Between Theoretical and Experimental Triaxial Behavior for Dr=60%

where the plastic modulus upon stress reversal K_p^* is taken to be proportional to the value of K_b on the p−axis (i.e., at $p = p_0$, Fig. 1), with the constant of proportionality being h_r, a model parameter. Note that M_t is the slope of $T = 0$ and η^f is the stress ratio normalized with respect to M_f. Thus M_t varies from ∞ on the isotropic axis to M_{t0} as the stress ratio becomes equal to M_f. M_{t0} is allowed to vary as a function of the "past maximum stress ratio η^{max}", with a model parameter α_r controlling its behavior.

The three-dimensional behavior is simulated by allowing some of the model parameters to vary smoothly between their values in triaxial compression and triaxial extension as a function of Lode angle.

Summary of the Salient Features of the Constitutive Model

The model is based on associated flow rule, and consequently produces a symmetric stiffness matrix. The model is capable of allowing plastic behavior during loading as well as during stress reversal, and therefore the pore pressure

Table 1: Model Parameters for Nevada Sands of Dr=40% (N-Sand40)

Dr (%)	λ^*	M_{fc}	G_0	n_e	m_1	h_0	h_1	h_r	α_r
40	.014	1.354	90	1.0	0.2	10.	8	1.0	.01
55	.011	1.450	90	1.0	0.2	25.	8	1.0	.01
60	.010	1.500	90	1.0	0.2	30.	8	1.0	.01
80	.007	1.700	90	1.0	0.2	100	8	1.0	.01

build-up for loading and stress reversals. Through the use of inner and outer surfaces, the phase transformation line is allowed to vary smoothly from $\alpha_g M_f$ to M_f as the loading progresses, with the line coinciding with the critical state line at failure. The particular choice of the deviatoric hardening function β allows the behavior to be simulated for a wide range of relative density of the sand, and for a wide range of effective mean normal pressure p with a single model.

Thus the model as presented above is intended to be quite general in the sense of being able to capture finer details of experimentally observed laboratory liquefaction behavior of sands. In the event of limited amount of data, some of the above parameters may be assigned fixed values.

In the finite element predictions presented here, a simplified version of the model, which requires only nine parameters, was used. In the model, the deviatoric hardening was ignored, and $g = 0$ and $f = 0$ were assumed to coincide. The model parameters are listed in Table 1, where the parameters for the relative density Dr=60% were obtained by calibration of the model against triaxial data (Anandarajah, 1993). In addition, the parameters were obtained for Dr=40% and 70% in Anandarajah (1993). From these, the parameters for Dr=55% and 80% were obtained by interpolation.

The parameters G_0 and n_e control the value of the shear modulus G according to (Richart et al., 1970):

$$G = G_0 p_a \frac{(2.97 - e)^2}{1 + e} (\frac{p}{p_a})^{n_e} \tag{12}$$

where p_a is the atmospheric pressure. m_1, h_0 and h_1 control the shape hardening function, h_r and α_r control the behavior upon stress reversal, and λ^* controls the density hardening function. A example of comparison between model simulations and triaxial behavior is presented in Fig. 2 for Dr=60%.

Finite Element Formulation

The numerical procedure is based on a set of fully-coupled dynamic equations for two-phase porous media. The finite element formulation employed is that of Zienkiewicz and Shiomi (1984). Specifically, the most general formulation, where the final matrix equations are in terms of solid and fluid dis-

Table 2: Parameters used in analysing Model 1 by HOPDYNE

k m/sec	n	ρ_s kN-sec^2/m^4	ρ_f kN-sec^2/m^4	K_0	γ' kN/m^3
3.3×10^{-3}	.420	2.7	1.0	.55	9.55

placements and pore pressure, is employed. The matrices involved in this form are symmetric. As indicated in the preceding section, the constitutive model yields a symmetric stiffness matrix. The entire analysis, therefore, involves symmetric global stiffness matrices.

In addition to the constitutive model parameters, the fully-coupled analysis requires the following properties of the soil: permeability (k), porosity (n), and density of solid (ρ_s) and fluid (ρ_f). The coefficient of earth pressure at rest (K_0) and the effective density (γ') are required in establishing the initial stresses in the soil. These properties are listed in Table 2 for Nevada sand for Dr=40%.

Computer Program HOPDYNE

HOPDYNE is a computer program to analyse two- and three-dimensional static and dynamic soil and soil-structure interaction problems by the finite element method. The program has the capability to model the soil, structural elements and soil/soil and soil/structure interfaces. The soil may be analysed either under drained, undrained or fully coupled condition. The latter allows simultaneous pore pressure build up and dissipation to be simulated, which is suitable for liquefaction analysis. Bounding surface based, isotropic and anisotropic, elasto-plastic models are used to represent the stress-strain behavior of clays (Dafalias and Herrmann, 1986; Anandarajah and Dafalias, 1986), and the two-surface model described in this paper to represent the behavior of sands (Anandarajah, 1994a; 1994b). The models may either be used in their most general forms or in their simplified forms involving a few parameters. The model parameters may be obtained either from a set of triaxial data or a combination of laboratory data and in situ data such as the pressuremeter, shear wave velocity and electrical data (Anandarajah and Agarwal, 1991; Anandarajah et al., 1986).

Eight and four noded 2-D isoparametric elements are available for modeling the soil in 2-D analyses, and twenty and eight noded 3-D elements are available for modeling in 3-D analyses. There are special slip elements to go with each type of soil elements. The slip elements employ relative displacements to avoid numerical problems. Soil and slip elements have the capability to model flow conditions. Space frame, bending elements are currently used to model structural elements such as piles, beams, columns, etc.

The type of loading that may be applied include earthquake loading and

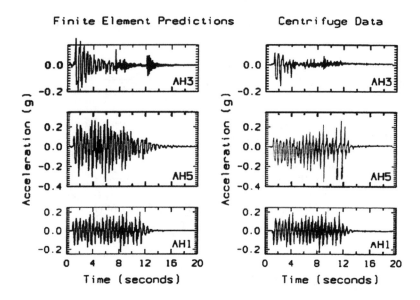

Figure 3: Comparison Between Theoretical and Experimental Accelerations for the Free-Field Problem Involving Liquefaction

dynamic loading acting anywhere within the domain. Static loads in the form of point loads and body forces may also be handled. This feature helps not only to analyse structures under monotonic static loads, but to establish initial stresses required for subsequent elasto-plastic analyses. Examples of analyses that may be performed by using HOPDYNE are: (1) Analysis of super structures (consisting of beam/column bending elements) subjected to machine vibration loads or earthquake loads applied at the base (e.g., pile cap level), (2) analysis of one-dimensional site response problems during earthquakes, (3) analysis of dams and slopes during earthquakes, (4) analysis of soil-structure interaction problems under machine vibration or earthquake loads (e.g., soil-pile-freeway interaction), etc.

HOPDYNE's capabilities have recently been further expanded by incorporating equivalent linear and p-y analysis capabilities. In addition, special radiation boundaries that can be used in nonlinear analyses (Anandarajah, 1990; 1993) are currently being incorporated.

Verification of the Validity of HOPDYNE

Recently, the validity of the predictions obtained using HOPDYNE has been verified for an earthquake soil-stucture interaction problem involving a

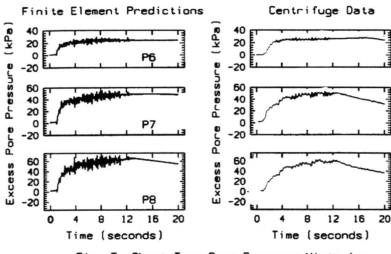

Fig. 5. Short Term Pore Pressure Histories.

Figure 4: Comparison Between Theoretical and Experimental Pore Pressures for the Free-Field Problem Involving Liquefaction

pile-soil-freeway system tested in a centrifuge. The superstructure was supported by two columns, each of which was supported by a group of piles founded on soft clay. A suitable two-dimensional model was used to model the problem. It was shown that the numerical results were reasonably close to the centrifuge observations (Anandarajah et al., 1995).

HOPDYNE was subsequently used in an a priori prediction exercise as part of the recent Velacs project. The behavior of three of the models tested in centrifuges were predicted; the numerical results are compared with experimental data here for one of the models. All of the problems involved liquefaction. The problem considered here is basically a free field site response problem, involving a 10-meter thick horizontal layer of uniform Nevada sand of 40% relative density; full details of the test may be found in Taboada and Dobry (1993). The problem was analysed using a column of eight-noded 2D finite elements. The model parameter set No. 1 listed in Table 1, and the soil properties listed in Table 2 were employed.

Compared in Fig. 3 are horizontal acceleration histories at the base (AH1), 5m above the base (AH5) and the ground (AH3). It is seen that the results in general compare reasonably well. Presented in Fig. 4 are comparisons of the pore water pressure time histories at heights of 2.5m (P8), 5m (P7) and 7.5m (P6) above the base. Again the results compare rather well. Note that as the pore water pressure builds up, the soil looses the ability to transmit the motion upward, and it is reflected on the recorded ground motion; HOPDYNE

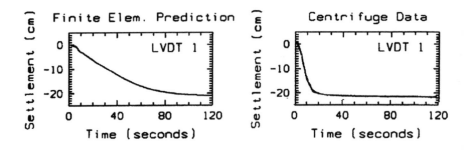

Figure 5: Comparison Between Theoretical and Experimental Settlement for the Free-Field Problem Involving Liquefaction

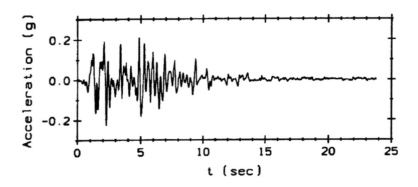

Figure 6: Base Motion Acceleration Used in Soil-Pile Interaction Analysis

is able to capture this aspect reasonably well. The predicted and observed ground settlements are compared in Fig. 5. The final value of the settlement is predicted well, however, the rate of settlement is under predicted. This indicates that the constitutive model represents the soil's behavior well, but the properties representing the dissipation have not been chosen right.

Finite Element Simulation of a Pile-Soil Interaction Problem

Here HOPDYNE is used to analyse one of the problems tested in a centrifuge by Wilson (Wilson, 1998; Boulanger, et al., 1999). The centrifuge model of interest here consisted on two horizontal layers of Nevada sand containing a number of model structures, including single piles and pile groups; here the behavior of a single pile is analysed. The bottom and top layers

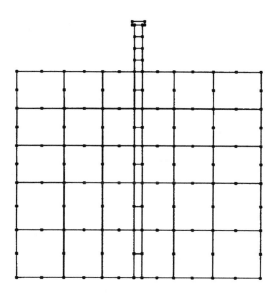

Figure 7: Finite Element Mesh Used for the Soil-Pile Interaction Analysis

were 11.4m thick dense Nevada sand (Dr=80%) and 9.3m thick medium dense (Dr=60%) Nevada sand respectively. The model is referred to as Csp3 and the reader may find the full details in Wilson (1998). Here the structure in prototype dimensions are considered for analysis. The pile has a diameter of 0.667m and a wall thickness of 72cm. The depth of embedment is 16.8m. The free length of the pile above the ground level is 3.81m. A weight of 500 kN was placed at the top of the pile. The cross sectional area, second moment of inertia and Young's modulus of the pile are 0.135 m^2, 0.006 m^4, and 70,000 kPa respectively. An equivalent, one-foot thick domain was used in the 2-D, plane strain finite element analysis. The values of momemt of inertia and the cross sectional area were increased by a factor of 1/0.667=1.5; that is, the pile is assumed to extend all the way in the z-direction, and a one-foot thick segment is chosen for the 2-D analysis. Therefore, the influence of that part of the soil in the z-direction that does not contain the pile is ignored.

The test where the pile was subjected to a scaled version of Kobe event (Event J) was considered. To save on the computer CPU time, the recorded motion was truncated before and after a certain time periods and the 22.5-second long middle strong motion shown in Fig. 6 was considered. Neglecting the influence of the soil below the pile tip, the 2D mesh shown in Fig. 7 was used in the analysis. The interactions between the pile and the adjacent structures and the bucket walls are also ignored. The free-field conditions were assumed to prevail at a distance of 10m sideways from the pile.

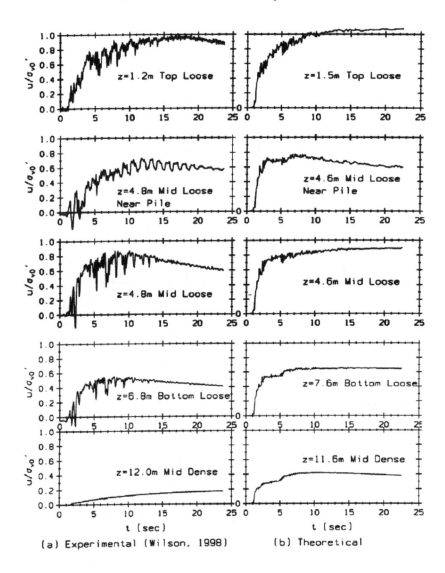

(a) Experimental (Wilson, 1998)   (b) Theoretical

Figure 8: Comparison Between Numerical and Measured Pore Water Pressures for the Soil-Pile Interaction Problem

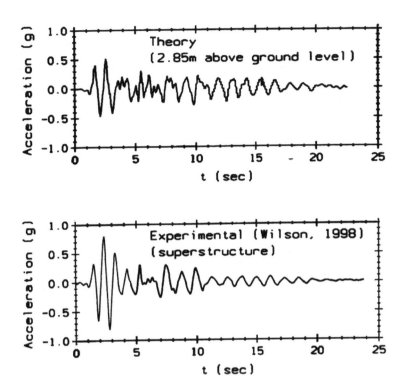

Figure 9: Comparison Between Numerical and Measured Accelerations for the Soil-Pile Interaction Problem

Presented in Fig. 8 are comparisons between numerical and experimental data on pore water pressure histories at several points away from the piles (on the far right vertical column, in the case of numerical results) and one point near the pile. The results are reasonably close. Measured acceleration time history of the superstructure is compared with that computed on the pile at 2.85m above the ground level in Fig. 9. While the overall agreement is encouraging, some level of discrepancy is to be noted. Further improvement in numerical predictions would require a three-dimensional analysis with consideration to interface behavior, which is currently in progress.

Conclusions

A two-surface elasto-plastic model capable of describing the stress-strain behavior of liquefiable sands is briefly described. This model has been implemented into a fully-coupled finite element analysis computer program HOP-

DYNE. The validity of HOPDYNE was verified in two recent projects, one involving clays and another involving liquefying soils. Numerical predictions compared well with experimental data in both cases. In the present paper, the earthquake behavior of a single pile measured in a centrifuge, reported in the literature, was numerically simulated using HOPDYNE and the results compared. The pile was founded on a layered saturated sandy deposit consisting of a dense and a medium dense sand layers. Both sand layers liquefied during the loading. The computed values of pore water pressures in the soil and the acceleration of the pile compared well with the measured data.

References

[1] Anandarajah, A. (1994a). "A Constitutive Model for Granular Materials Based on Associated Flow Rule." *Soils and Foundations*, Japanese Society of Soil Mechanics and Foundation Engineering, 34(3):81-98.

[2] Anandarajah, A. (1994b). "Procedures for Elasto-Plastic Liquefaction Modeling of Sands." *Journal of Engineering Mechanics Division*, ASCE, 120(7):1563-1589.

[3] Anandarajah, A., Rashidi, H. and Arulanandan, K. (1995). "Elasto-Plastic Finite Element Analyses of Earthquake Pile-Soil-Structure Interaction Problems Tested in a Centrifuge." *Computers and Geotechnics*, 17:301-325.

[4] Anandarajah, A. (1993). "Dynamic analysis of Axially-Loaded Footings in Time Domain." *Soils and Foundations*, Japanese Society of Soil Mechanics and Foundation Engineering, 33(1):40-54.

[5] Anandarajah, A., and Agarwal, D. (1991). "Computer-Aided Calibration of a Soil Plasticity Model." *International Journal for Numerical and Analytical Methods in Geomechanics*, 15(12):835-856.

[6] Anandarajah, A. (1990). "Time-Domain Radiation Boundary for Analysis of Plane Love-Wave Propagation Problems." *International Journal of Numerical Methods in Engineering*, 29:1049-1063.

[7] Anandarajah, A. and Dafalias, Y.F. (1986). "Bounding Surface Plasticity, Part 3: Application to Anisotropic Soils." *Journal of Engineering Mechanics Division*, 112(12):1292-1318.

[8] Anandarajah, A. (1993). "VELACS Project: Elasto-Plastic Finite Element Prediction of the Liquefaction Behavior of Centrifuge Models Nos. 1, 3 and 4a," *Proc. Intl. Conf. on Verification of Numerical Procedures for the Analysis of Soil Liquefaction Problems*, Davis, California, Oct. 17-20, eds. K. Arulanandan and R. F. Scott, 1, 1075-1104.

[9] Anandarajah, A. (1990). "HOPDYNE – A Finite Element Computer Program for the Analysis of Static, Dynamic and Earthquake Soil and Soil-Structure Systems" The Johns Hopkins University, Baltimore, Maryland.

[10] Anandarajah, A., Meegoda, N.J., and K. Arulanandan, K. (1986). "In Situ Stress-Strain Prediction of Fine Grained Soils." *Proceedings of ASCE specialty conference on "Use of In Situ Testing in Geotechnical Engineering,"* held in June 22-25 at Virginia Tech, Blacksburg, Virginia.

[11] Boulanger, R.W., Wilson, D.W., Kutter, B.L. and Abghari, A. (1999). "Soil-Pile-Superstructure Interaction in Liquefiable Sand." *Transportation Research Record* 1569, pp. 55-64.

[12] Castro, G. (1969). "Liquefaction of Sands," Harvard SM Series No. 81.

[13] Dafalias, Y. F. and Herrmann, L. (1986). "Bounding Surface Plasticity. II: Application to Isotropic Cohesive Clays." *Journal of Engineering Mechanics*, ASCE, 112(12):1242-1291.

[14] Richart, F. E., Hall, J. R., Jr., and Woods, R. D. (1970). "Vibrations of Soils and Foundations," Prentice-Hall, Inc., Inglewood Cliffs, New Jersey, 1970.

[15] Schofield, A., and Wroth, P. (1968). Critical State Soil Mechanics, Mc-Graw Hill publication.

[16] Taboada, V. M. and Dobry, R. (1993). "Experimental Results of Model 1 at RPI," *Proc. Intl. Conf. on Verification of Numerical Procedures for the Analysis of Soil Liquefaction Problems*, Davis, California, Oct. 17-20, eds. K. Arulanandan and R. F. Scott, 1, 3-18.

[17] Wilson, D. W. (1998). "Soil-Pile Superstructure Interaction in Liquefiable Sand and Soft Clay." Doctoral Dissertation, University of California, Davis, 1998, 173pp.

[18] Zienkiewicz, O. C. and Shiomi, T. (1984). "Dynamic Behavior of Saturated Porous Media: The Generalized Biot Formulation and Its Numerical Solution." *International Journal for Numerical and Analytical Methods in Geomechanics*, 8:71-96.

EVALUATION OF DYNAMIC SITE RESPONSE AND DEFORMATION OF INSTRUMENTED SITE - KOBE

K. Arulanandan [1] M. ASCE, X. S. Li [2] M. ASCE and K. (Siva) Sivathasan [3] S. M. ASCE

ABSTRACT: The observed dynamic response of an instrumented site at Port Island during the 1995 Kobe earthquake was utilized to demonstrate the feasibility of computer simulation of earthquake induced site response and liquefaction-induced deformations of a level ground site. Non-destructive in-situ electrical and shear wave velocity methods were used to obtain the initial state parameters and constitutive model constants representative of the site. The analysis used the fully coupled, effective stress-based, nonlinear, finite element program SUMDES with a reduced order bounding surface hypo-plasticity model to simulate the stress-strain behavior of cohesive soils and modified reduced order bounding surface hypo-plasticity model to simulate the stress-strain behavior of non-cohesive soils. The results of the dynamic analysis such as acceleration time histories and liquefaction-induced deformations agreed reasonably well with the acceleration time histories and liquefaction-induced vertical and horizontal deformation behaviors observed during the Kobe earthquake. The results of this study show that computer simulation of earthquake effects at level ground sites is possible using non-destructive in-situ testing and a verified numerical procedure.

INTRODUCTION

Before the event, evaluation of the response of our infrastructure systems subjected to earthquakes is a desirable objective for the adaptation of appropriate, safe and economical retrofit measures of existing facilities and safe and economical design of new infrastructure

---

[1] Professor, Dept. of Civ. and Envir. Engrg., Univ. of California, Davis, CA 95616.
[2] Assistant Professor, Hong Kong University of Science and Technology, Hong Kong, China.
[3] Graduate Student, Dept. of Civ. and Envir. Engrg., Univ. of California, Davis, CA 95616.

facilities. Once a verified numerical procedure and in-situ properties representative of the site are available, computer simulation of earthquake effects of any site can be carried out. The objective of this paper is to demonstrate how computer simulation of earthquake effects such as site response and liquefaction-induced deformation of an instrumented site at Port Island, Kobe can be achieved using in-situ properties representative of the site.

EARTHQUAKE AND SOIL CHARACTERISTICS AT INSTRUMENTED SITE

On January 17, 1995, at 5:46 A.M. an earthquake of magnitude 7.2 (Richter scale) struck the southern part of Hyogo prefecture, caused severe damage to the infrastructure. The location of the seismographs at Port Island recording station and shear wave velocity profile are shown in Fig. 1 (Toki 1995). The soil profile at an instrumented site in Port Island, Kobe consists of a reclaimed loose surface layer down to about a 18m depth; a silty clay layer between 18 and 28m in depth; gravelly sand and silt layers between 28 and 38m in depth; alternate layers of gravelly sand and silt in between 38 and 60m depth; a diluvial silty clay layer between 60 and 83m in depth; and sand with gravel layers inter-layered with clay starting at about 83m depth. The water table is located at approximately a 3m depth. A representative shear wave velocity profile and parameters obtained from non-destructive electrical method are shown in Fig. 2. Accelerometers were placed at four different depths. At each depth three different directional components of accelerations were reported. The corrected-recorded accelerations at different depths in different directions are presented in Fig. 3a (Gifu University). Fig. 3a shows that the strong shaking phase at Port Island, Kobe lasted for about 10s (from 13 to 23s). Soil liquefaction at shallower depths appears to have precluded high acceleration from occurring at the ground surface.

VERIFIED NUMERICAL PROCEDURE SUMDES

SUMDES is a computer program to perform dynamic response analyses of $\underline{S}$ites $\underline{U}$nder $\underline{M}$ulti-$\underline{D}$irectional $\underline{E}$arthquake $\underline{S}$haking (Li et al. 1992). The basic assumptions made in the formulation are listed below:

1. Site is horizontally layered and extends infinitely in horizontal directions,

2. Ground surface is free of stresses,

3. Water flow is laminar and Darcy's law holds,

4. Bottom boundary is impermeable,

5. Soil layers below water table are fully saturated, and

6. Waves travel along vertical direction only

The procedure is formulated on the basis of effective stress principle, vectored motion, transient pore fluid movement, and generalized material stiffness; therefore, it is capable of predicting three-directional motions, the pore pressure buildup and dissipation within the soil deposits. The overall procedure consists of two computer programs: SUMDES and TESTMODL. SUMDES performs site response analyses; TESTMODL is used to examine the responses of the built-in inelastic models with given model parameters and loading conditions. This numerical procedure was verified during the VELACS project. Numerous examples have been published that demonstrate the applicability of the methodology which utilizes numerical procedure SUMDES and non-destructive in situ electrical method of site characterization (Reyes et al. 1996; Arulanandan et al. 1996; Sivathasan et al. 1998; Arulanandan and Sivathasan 1998).

CONSTITUTIVE MODELS USED IN THIS ANALYSIS

1. Reduced Order Bounding Surface Hypo-plasticity Model:

The reduced order bounding surface hypo-plasticity model (Li et al. 1992, Li 1996) is a special version of the original bounding surface hypo-plasticity model (Wang et al. 1990). The model has a smaller number of model parameters than the original model (due to the special loading conditions imposed by site response analysis), but still retains the capability to simulate liquefaction potential and pore water pressure generation. In the reduced order bounding surface plasticity model, ten parameters (see Table 1) are involved in the site response analysis. Among the ten, three parameters (b, h_p and R_p/R_f) are either inactive in the given loading conditions or vary within a small range, so they can be assumed empirically. Two parameters d and k_r can be calibrated based on cyclic strength data obtained from field measurement. Other parameters can be obtained from established correlations between electrical parameters and required parameters.

2. Modified Reduced Order Bounding Surface Hypo-plasticity Model:

The reduced order hypo-plasticity model is modified according to a recent paper by Li et al. (1999) to improve the capability in modeling the constitutive behavior at high strains and most importantly, to incorporate the critical state concept in the sand response. The modification consists of rendering the phase transformation line a function of the state parameter ψ, which measures the difference between the current and critical void ratios at same mean normal pressure, p, such that when the state parameter is zero, the phase transformation line becomes identical to the critical state line in q – p space. This idea was originally proposed and applied in another constitutive model by Manzari and Dafalias (1997). As a result of this modification, the dilatancy depends on the state in a way which yields a zero value at critical state. This dependence allows a realistic modeling of the response of a sand in either loose or dense state, or in the transition from one state to another state. After the modification, dilatancy is not only a function of stress state but also a function of internal material state.

Li et al. (1999) proposed the phase transformation stress ratio

$$R_p = R_f e^{(\text{sgn}\,\psi)m|\psi|^n} \tag{1}$$

It can be seen that according to Eq. 1 the phase transformation stress ratio R_p becomes an exponential function of the state parameter ψ for all material states.

Referring to the definition of the moduli K_r and H_r of the original hypo-plasticity model (Wang et al. 1990), the dilatancy d is given by

$$d = \sqrt{\frac{3}{2}} \frac{H_r}{K_r} = \sqrt{\frac{3}{2}} \frac{G}{K} \frac{h_r}{k_r} \left(\frac{p}{p_m}\right)^a \left(\frac{R_m}{R_f}\right)^b \left(\frac{\omega R_p - R_m}{R_m}\right) = d_0 \left(\frac{\omega R_p - R_m}{R_m}\right) \tag{2}$$

where $R = \sqrt{r_{ij} r_{ij} / 2}$ is a generalized stress ratio invariant; $r_{ij} = s_{ij} / p$ is the deviatoric stress ratio with s_{ij} the deviatoric stress tensor; p the mean normal pressure; ($R = \eta / \sqrt{3}$ for triaxial conditions and is the multi-axial counterpart of η; $\eta = q / p$ is the stress ratio), and R_p, R_m and R_f are the values of R for the phase transformation, the historically maximum, and ultimate failure, respectively. The p_m is the maximum p in the loading history; h_r and k_r are two model constants that control the plastic deviatoric and volumetric hardening respectively; a and b are two other model constants; and ω is a function of stress increment direction that introduces the dependence of phase transformation stress ratio on shearing direction change. As can be seen, since R_p is now a ψ−dependent variable, d becomes ψ−dependent.

In order to determine the state parameter, ψ, a critical state line in the e – p plane needs to be defined. It has been shown experimentally (eg. Verdugo and Ishihara 1996) that the critical state line for sand can not be approximated by a straight line in the e – p space if a wide range of pressure is considered. The following relationship (originally proposed by Wang et al. 1990 for the virgin compression line of sand) is adopted to describe this line (Li 1997):

$$e_c = e_\Gamma - \lambda_c \left(\frac{p_c}{p_a} \right)^\xi \tag{3}$$

where e_c and p_c are the critical void ratio and associated critical mean normal stress, respectively. p_a is a reference pressure usually set to equal the atmospheric pressure (101kPa) for convenience, and λ_c, ξ and e_Γ are dimensionless material constants. Locating the critical state line in the e – p space is of critical importance for the determination of the state parameter, ψ. It has been shown that the critical state line in the e - p space can be closely approximated by considering the relationship between the average shape factor and the position of the critical state line of sands based on laboratory test data. For the Port Island sand, the location of the critical state line in e - p space is shown in Fig. 4 (Arulanandan 1995; Arulanandan and Sivathasan 1995). To find model parameters λ_c, ξ and e_Γ, one has to transfer the critical state line to e - $\left(\dfrac{p}{p_a} \right)^\xi$ space as shown in Fig. 5.

The elastic shear and bulk moduli G and K are given by the following empirical equation (Wang et al. 1990).

$$G = G_o p_a \frac{(2.973 - e)^2}{1 + e} \sqrt{\frac{p}{p_a}} \tag{4}$$

$$K = p_a \frac{1 + e}{\kappa} \sqrt{\frac{p}{p_a}} \tag{5}$$

in terms of a parameter G_o for G and κ for K. Note that in the original model the initial void ratio e_o was used instead of e in the above equations. The modification was made with the aim of unifying the model parameters for all densities. Using the standard linear elasticity expression of $G/K = 3(1-2v)/2(1+v)$ and Eqs. 4 and 5, one can write the expression for κ as

$$\kappa = \frac{3(1-2v)}{2G_o(1+v)}\left[\frac{1+e}{2.973-e}\right]^2 \tag{6}$$

It can be shown from the formulation of the original hypo-plasticity model that the value of v only affects the drained response and can be calibrated from drained test results, independently of the undrained ones. Micro-mechanics investigations have shown that the Poisson's ratio of an assemblage of particles is significantly smaller than that of particles' material (Chang and Misra, 1990). The value of v falls into a relatively narrow range of 0 to 0.1.

Eq. 2 expresses the variation of dilatancy d, as derived from the general hypo-plasticity model formulation, in line with Eq. 1 which yields the variation of d with ψ. But a constant value of d_0 in Eq.2 would facilitate the calibration of the constants m and n. A process of fitting the stress-strain curves of the Port Island sand yielded the values $d_0 = -0.76$, m = 3.6 and n = 0.75.

This modified bounding surface hypo-plasticity model has only twelve model parameters but still maintains the capability to simulate liquefaction potential, pore water pressure generation and dissipation and the constitutive behavior at high strains for a wide range of material densities and confining pressures. The constitutive model parameters and state parameters in the modified reduced order bounding surface hypo-plasticity model are shown in Table 2.

NON-DESTRUCTIVE ELECTRICAL METHOD OF IN-SITU CHARACTERIZATION

Considerable research, for an example the VELACS project, has been undertaken to verify numerical procedures using laboratory test results and known initial state parameters. Difficulty in obtaining undisturbed samples and carrying out reliable laboratory tests to avoid disturbance of the sample to calibrate constitutive models seem to be a major obstacle in advancing the use of fully coupled, effective stress-based, nonlinear, numerical procedures using elastoplastic constitutive models. The in-situ non-destructive site characterization is necessary to get the initial state parameters and constitutive model constants representative of the site for the computer simulation of the site to earthquake.

It has been shown experimentally by numerous studies conducted by Archie (1942), Arulanandan and Kutter (1978) and Wyllie and Gregory (1957) that a non-dimensional electrical parameter, called the formation factor, is dependent upon the *particle shape, long axis orientation, contact orientation* and *size distribution*, and also *cementation, degree of saturation, void ratio* and *anisotropy*. The formation factor is thus a function of the fabric of the soil and varies with stress to the extent that the fabric is altered by the stress. An index, the formation factor F, is experimentally obtained as the ratio of the conductivity of the pore fluid and the conductivity of the soil sample. Thus, the formation factor is a non-dimensional parameter. It depends on sand structure, especially on its elements associated with particle arrangement, particle shape and porosity. The dependence of the formation factor on void ratio, particle shape and long axis orientation has been shown theoretically by Arulanandan and Dafalias (1979), and Dafalias and Arulanandan (1979). The formation factor F has been shown to be a valuable directional index parameter when the sand structure is anisotropic. Due to the above facts, in-situ measurements of formation factor (F) provides a means to characterize the structure of sands in the field and to obtain empirical correlations between shape and the anisotropic state of sand characterized in terms of F and the engineering behavior such as liquefaction characteristics of sands.

The formation factor can be used to quantify and predict the porosity, shape and anisotropy of sand deposits. An average formation factor $\overline{F}$, is given by

$$\overline{F} = \frac{(F_v + 2F_h)}{3} \tag{7}$$

where, F_v is the vertical formation factor and F_h is the horizontal formation factor.

It has been shown that $\overline{F}$ is independent of anisotropy caused by the orientation of preferred particles, and is a direct measure of porosity (n), for a given sand. For practical purposes, an integration technique proposed by Bruggeman (1935) was used by Dafalias and Arulanandan (1979) to derive an expression for the average formation factor, $\overline{F}$, as a function of porosity (n), and average shape factor ($\overline{f}$), as

$$\overline{F} = n^{-\overline{f}} \tag{8}$$

It has been shown both theoretically and experimentally that the shape factor is directional and depends on porosity, gradation and the particles' shape and orientation of the particles (Dafalias and Arulanandan 1979, Arulanandan and Kutter 1978). An anisotropy index (A), was introduced by Arulanandan and Kutter (1978) as

$$A = \left(\frac{F_v}{F_h} \right)^{0.5} \tag{9}$$

The anisotropy of sand structure is due to the orientation of individual particles and the contact orientation.

IN-SITU PROPERTIES USING NON-DESTRUCTIVE IN-SITU ELECTRICAL METHOD

1. ϕ – Friction angle

Using electrical measurements, one obtains $\overline{f}$,

the slope of the critical state line (M) in the q – p' space is obtained as

shown in the paper by Arulanandan et al. 1994.

Moreover, from

$$M = \frac{(6Sin\phi)}{(3 - Sin\phi)} \tag{10}$$

one obtains the critical state friction angle(ϕ).

2. n – Porosity

Using electrical measurements one obtains porosity using an established

relationship between formation factor and porosity for a particular soil

as shown in Arulanandan and Muraleetharan 1985 (see Fig. 6).

3. K_o – The earth pressure coefficient at rest

Using electrical measurements one obtains K_o from pre-established

correlation between the coefficient of earth pressure at rest and the

electrical index $A^4\overline{f}$ (Meegoda and Arulanandan 1986).

4. k – The coefficient of hydraulic conductivity

Specific surface area can be obtained from grain size distribution.

The SSA(Specific Surface Area) is defined as the surface area per unit

volume of soil particles, i.e.,

$$SSA = \frac{6[(W_1 / d_1) + ((W_2 - W_1)/ d_2) + \ldots\ldots\ldots + ((100 - W_{n-1})/ d_n)]}{100} \tag{11}$$

where, $W_1, W_2,$........ etc., are the percentage passing; and d_i is the average diameter between W_i and W_{i-1}.

By means of electrical measurements, one obtains the coefficient of hydraulic conductivity from pre-established correlation between vertical permeability of

sands and silty sands, and the electrical index defined by $\dfrac{\overline{F}^{\frac{-3}{f}}}{\left(1-\overline{F}^{\frac{-1}{f}}\right)^2 (SSA)^2}$

using Kozeny-Carman equation.

As shown by Arulanandan and Muraleetharan (1988), there is a correlation between k and the above electrical index parameter, which is approximately linear.

5. G_0 – This parameter is the modulus coefficient defining the initial (maximum) elastic shear modulus G_{max} using the following equation (Li et al. 1992):

$$G_{max} = \frac{G_o(2.973 - e_o)^2 (P \cdot P_{atm})^{0.5}}{(1 + e_o)} \qquad (12)$$

$$G_{max} = 1000 K_{2max} (P)^{0.5} \qquad (13)$$

K_{2max} depends on void ratio, strain amplitude geological age of the sand mass and in situ stresses. P is the mean effective confining pressure, in pounds per square foot and G_{max} is given in pound per square foot. A correlation between K_{2max} and the electrical parameter $\overline{F}/(A\overline{f})^{0.5}$ is given by Arulanandan et al. (1983) and is used to obtain K_{2max}.

6. κ – The slope of the rebound line in e – ln(p') space is defined by (Li et al. 1992)

$$\kappa = \frac{3(1 + e_o)^2 (1 - 2v)}{2G_o (2.973 - e_o)^2 (1 + v)} \tag{14}$$

7. h_r – The parameter which characterizes the relationship between shear modulus and shear strain magnitude. The value of h_r can be determined based on the standard modulus reduction curve (Li et al. 1992) or the secant shear modulus measured at a control shear strain. The parameter h_r so calibrated is in general a function of density (Li et al. 1999).

8. b – A parameter affecting the shape of the stress paths of the virgin shear loading. A typical value of b is 2.0.

9. k_r – The parameter which characterizes the amount of the effective mean normal stress change caused by shear loading. k_r value also affects the liquefaction resistance. Since d_0 is a constant k_r is not an independent parameter (Li et al. 1999). k_r must satisfy the calibrated value of d_0.

10. d– The parameter which characterizes the rate of the effective mean normal stress change caused by shear unloading. This parameter is the major one controlling the liquefaction resistance of a material.

PARAMETER BASED ON CYCLIC STRENGTH DATA (d)

The cyclic strength is estimated based on in situ electrical method (Arulmoli et al. 1985). For example, cyclic strength curve for electrical parameter $A^3 /(\overline{Ff}) = 0.26$ is shown on Fig. 7. This curve can be simulated by running the utility program TESTMODL from the program package of SUMDES for a single element response with other calibrated parameters control the development of pore pressure under undrained conditions.

The calibration procedure to find d is a process of trial and error:

1. Select cyclic stress ratio, with a specified mean normal confining stress;

2. Assume a value of d;

3. Run the utility program TESTMODL for the selected stress level and count the number of cycles to cause liquefaction (see Fig. 8);

4. Compare the results obtained from step 3 against the given relationship of stress ratio required to cause liquefaction versus number of cycles to cause liquefaction. Adjust the d value if needed;

5. Repeat steps 3 and 4 until satisfactory result is obtained.

Repeat same procedure for two or more shear stress amplitudes, with a specified mean normal stress.

METHOD OF ANALYSIS

Samples were taken from the man-made fill at the instrumented site at Port Island, Kobe and correlation between formation factor and porosity were estabilished at the University of California at Davis soil mechanics laboratory (see Fig. 6). From the non-destructive electrical measurements, average shape factor and anisotropy index were calculated. Due to unavailability of samples from the bottom sand layers, the same average shape factor and anisotropy index were assumed for the bottom sand layers. Using shear wave velocity measurements, the maximum shear modulus was calculated. The average formation factor was obtained using the maximum shear modulus, average shape factor and anisotropy index. Porosity was obtained from pre-estabilished relationship (see Fig. 6) between formation factor and porosity using estimated average formation factor. All the initial state parameters and constitutive model constants were calculated using the average formation factor, average shape factor and anisotropy index.

ANALYSIS RESULTS

The numerical simulation of the instrumented site at Port Island, Kobe was performed by the computer program SUMDES. In the analysis, multi-directional shaking was applied to the soil deposits. The input motions and their response spectrum are shown in Fig. 9. The reduced order bounding surface hypo-plasticity model was used to simulate the cohesive soil behavior and the modified reduced order bounding surface hypo-plasticity model was used to simulate the non-cohesive soil behavior. Calibrated constitutive model constants and state parameters are provided in the Tables. 1 and 2.

From the specification for the instrumented site at Port Island, Kobe, accelerometers were placed at four different depths. At each depth three different directional components of acceleration were reported. In the present analysis, all the predictions and comparisons were made at the four different depths in three different directions.

Evaluated accelerations at different depths in different directions are presented in Fig. 3b. Response spectra from analysis results and recorded accelerations at the ground surface are compared in Fig. 10. Note that recorded and SUMDES results compare favorably. Evaluated excess pore water pressure at different depths are shown in Fig. 11. No field pore water pressure measurements were reported for comparison with the evaluated pore water pressure during the earthquake shaking. The pore water pressure ratio is a measure of liquefaction potential. The maximum pore water pressure ratio variation with depth is shown in Fig. 12. It shows that the zone of high pore water pressure ratio in excess of 90% extended in the first 18m from ground surface and at a depth of 30m. Vertical settlement variation with time at the ground surface is shown in Fig. 13. Reported average vertical settlement at the ground surface is as high as 40-50cm (Ishihara et al. 1996). Evaluated lateral displacements variation with depth in two different directions are shown in Fig. 14. Reported maximum lateral displacement 1~3m occurred at the waterfront (Ishihara et al. 1996). The observed lateral displacements were generally large near the waterfront and decreased with distance inland. These results provide reliability to the computational analysis.

CONCLUSIONS

A method of determining the in-situ porosity and the position of the critical state line in the e - p space using the average shape factor to determine the in-situ state parameter, ψ, is presented.

The process of pore water pressure generation, the development of liquefaction-induced deformations were simulated using the nonlinear, fully coupled, effective stress analysis. The accelerations, calculated from the analysis, are in close agreement with the recorded accelerations on the Port Island down-hole array. This analysis shows liquefaction occurred in the man-made fill and agrees with observed liquefaction phenomenon at the Port Island, Kobe. Evaluated vertical settlement and lateral displacements are in close agreement with the observed vertical settlement and lateral displacements at the Port Island area.

It has been demonstrated that it is possible to evaluate the dynamic response of a level ground instrumented site using verified numerical procedure and input properties such as initial state parameters (porosity, coefficient of earth pressure at rest, hydraulic conductivity) and constitutive model constants (slope of the rebound line, friction angle, maximum shear modulus, slope of the critical state line in e_c - p space, hardening parameters, etc.) representative of the site obtained by non-destructive in situ testing.

ACKNOWLEDGMENT

The writers are indebted to Prof. Y. F. Dafalias for reviewing the manuscript and providing valuable comments.

Appendix I. References

Archie, G. E. (1942). "The Electrical Resistivity Log as an Aid in Determining Some Reservoir Characteristics," Transactions, American Institute of Mining, Metallurgical and Petroleum Engineers, Vol. 146, pp.54-61.

Arulanandan, K. (1995). Lecture on In-Situ Soil Characterization, Location of Critical Void Ratio Vs Mean Normal Stress by In-Situ Testing, Kyoto University, Kyoto, Japan.

Arulanandan, K., Anandarajah, A., and Meegoda, N. J. (1983). "Soil Characterization for Non-Destructive In-Situ Testing," Symposium Proceedings part 2, The interaction of Non-Nuclear Munitions with Structures, US Air Force Academy, Colorado, May 10 – 13.

Arulanandan , K., and Dafalias, Y. F. (1979). "Significance of Formation Factor in Sand Structure Characterization," Letters in application and Engineering Sciences, Vol. 17, pp.109-112.

Arulanandan, K. and Kutter, B. L. (1978). "Directional Structure Index Related to Sand Liquefaction," Proceedings of the June 19-21, Specialty Conference on Earthquake Engineering and Soil Dynamics, ASCE, Pasadena, California, pp. 213-229.

Arulanandan, K., Li, X. S., Paulino, G. H. and Sivathasan, K. (1996). "Dynamic Response of Saturated Level Ground Sites using Verified Numerical Procedure and In-Situ Testing," Proceedings of the National Science Foundation and California Transportation Dept. Sponsored Workshop/Conf. on Application of Numerical Procedures in Geotech. Earthquake Engrg., Univ. of California, Davis.

Arulanandan, K. and Muraleetharan, K. K. (1985). "Soil Liquefaction - A Boundary Value Problem (A Priori Prediction of Pore Pressure Generation and Dissipation During Earthquakes on Level Ground Sites," Technical Report, University of California, Davis.

Arulanandan, K. and Muraleetharan, K. K. (1988). "Level Ground Soil-Liquefaction Analysis using In-situ properties: I," Journal of Geotechnical Engineering, ASCE, Vol. 114, No. 7.

Arulanandan, K. and Sivathasan, K. (1995). "In-Situ Prediction of Critical Void Ratio Vs Mean Normal Pressure," Technical Report, University of California, Davis.

Arulanandan, K. and Sivathasan, K. (1998). "Evaluation of Site Response and Deformation of Instrumented Bridge Sites Subjected to Large Magnitude Earthquakes," Preliminary Report to Dept. of Transportation, California.

Arulanandan, K., Yogachandran, C and Hossain Rashidi (1994). "Dielectric Dispersion and Formation Factor Methods of Site Characterization," Geophysical Characterization of Sites, Volume prepared by the ISSME Technical Committee for the XIII International Conference on Soil Mechanics and Foundations Engineering, New Delhi, India.

Arulmoli, K., Arulanandan, K. and Seed, H. B. (1985). "New Method for Evaluating Liquefaction Potential," Journal Geotechnical Engineering Division, ASCE, 111(1), 95-114.

Bruggeman, D. A. G. (1935). "Berechung Verschiedenez Physikalischer Konstanten Von Heterogenen Substanzen," Ann. Phys. Lpz. 5, Vol. 24, pp. 636.

Chang, C. S. and Misra, A. (1990). "Packing Structure and Mechanical Properties of Granulates," J. of Engineering Mechanics, ASCE, Vol. 116, No. 5, pp. 1077-1093.

Dafalias, Y. F. and Arulanandan, K. (1979). "Electrical Characterization of Transversely Isotropic Sands," Archives 0f Mechanics, Warsaw, 31(5), 723-739.

Ishihara, K., Yasuda, S. and Nagase, H. (1996). "Soil Characteristics and Ground Damage," Soils and Foundations, Special Issue on Geotechnical Aspects of the January 17, 1995 Hyogoken-Nambu Earthquake, pp. 109-118, January.

Li, X. S. (1996)." Reduced-Order Sand Model for Ground Response Analysis," Journal of Engineering Mechanics, ASCE, 122(9), pp. 872-881.

Li, X. S. (1997). "Modeling of Dilative Shear Failure," Journal of Geotechnical Engineering, ASCE, Vol. 123, No. 7, pp.609-616.

Li, X. S., Dafalias, Y. F. and Wang, Z. L. (1999). "A Critical-State Hypo-plasticity Sand Model with State Dependent Dilatancy," Canadian Geotechnical Journal (accepted).

Li, X. S., Wang, Z. L., and Shen, C. K. (1992). "SUMDES: A Non Linear Procedure for Response Analysis of Horizontal-Layered Sites Subjected to Multi-Directional Earthquake Loading," Department of Civil and Environmental Engineering, University of California at Davis.

Manzari, M. T. and Dafalias, Y. F. (1997). "A Critical State Two-Surface Plasticity Model for Sands," Geotechnique, Vol. 47, No. 2, pp. 255-272.

Meegoda, N.J. and Arulanandan,K. (1986). "Electrical Method of Predicting In-Situ Stress State of Normally Consolidated Clays," Proceedings of the In-Situ 86', ASCE Specialty Conference, Blacksburg, VA, pp794-808.

Reyes, C., Arulanandan, K., Steve Mahnke, Chad Baker and Sivathasan, K. (1996). "Fully Coupled Effective Stress Based Analysis to Investigate the Consequences of Soil Liquefaction at Mosher Slough for FEMA Project," Report to Kleinfelder and Associates, Stockton, California.

Sivathasan, K., Paulino, G. H., Li, X. S. and Arulanandan, K. (1998). "Validation of Site Characterization Method for the Study of Dynamic Pore Pressure Response," Geotechnical Special Publication No. 75, Volume one, Geotechnical Earthquake Engineering and Soil Dynamics III, ASCE, Seattle, Washington.

Toki, K. (1995). Committee of Earthquake Observation and Research in the Kansai Area.

Verdugo, R. and Ishihara, K. (1996). "The Steady State of Sandy Soils," Soils and Foundations, Vol. 36, No. 2, pp.81-91.

Wang, Z.L., Dafalias, Y.F. and Shen, C.K. (1990). "Bounding Surface Hypo-plasticity Model for Sand," Journal of Engineering Mechanics, ASCE, Vol.116, No. 5, May, pp983-1001.

Wyllie, M. R. J. and Gregory, A. R. (1953). "Formation Factors of Unconsolidated Porous Media," Influence of Particle Shape and Effect of Cementation, Petroleum Transactions, American Institute of Mining, Metallurgical and Petroleum Engineers, Vol. 198, pp.103-109.

Appendix II. Notation

The following symbols are used in this paper:

A	= anisotropy index
a	= model constant
b	= model constant
b	= parameter affecting the shape of the stress paths of the virgin shear loading
d	= parameter which characterizes the rate of the effective mean normal stress change caused by shear unloading
d	= dilatancy
d_o	= constant related to dilatancy
$d_1, d_2,...d_n$	= particle size
e	= 2.718281828
e	= void ratio
e_c	= critical void ratio
e_o	= initial void ratio
e_Γ	= void ratio at unit pressure in $e - \left(\dfrac{p}{p_a}\right)^\xi$ space
$\bar{F}$	= average formation factor
F_h	= horizontal formation factor
F_v	= vertical formation factor
$\bar{f}$	= average shape factor
G	= elastic shear modulus
G_{max}	= maximum shear modulus
G_o	= constant related to shear modulus

H_r = plastic shear modulus

h_p = parameter controlling the amount of shear strain increment due to the change of the maximum effective mean normal stress in the loading history

h_r = parameter which characterizes the relationship between shear modulus and shear strain magnitude

K = elastic Bulk modulus

K_o = coefficient of earth pressure at rest

K_{2max} = modulus related to maximum shear modulus

K_r = plastic bulk modulus

k = coefficient of hydraulic conductivity

k_r = parameter which characterizes the amount of the effective mean normal stress change caused by shear loading

M = slope of the critical state line in q-p space

m = model constant

n = model constant

n = porosity

p = mean normal stress

P_a = reference pressure

P_{atm} = atmospheric pressure

P_c = critical mean normal stress

p_m = historically.maximum mean normal pressure

q = deviatoric stress

R_f = stress ratio at failure state

R_m = historically maximum stress ratio

R_p = stress ratio at phase transformation state

R_p/R_f = ratio of stress ratio at phase transformation state and stress ratio at failure state

r_{ij} = deviatoric stress ratio

SSA   = specific surface area

$sgn(\psi)$ = sign of state parameter

s_{ij} = deviatoric stress tensor

$W_1, W_2,..W_n$ = weight percentage passing through the sieve

ϕ = friction angle

η = stress ratio

κ = slope of the rebound line in e-ln(p) space

λ_c = slope of the critical state line in $e - \left(\dfrac{p}{p_a}\right)^{\xi}$ space

ν = Poisson's ratio

ω = function of stress increment direction that introduces the dependence of phase transformation stress ratio on shearing direction change

ξ = constant

ψ = state parameter

Subscripts

i, j   = positive integer indices

p   = phase transformation

f   = failure

c   = critical

atm   = atmospheric

max   = maximum

o   = initial

h   = horizontal

v   = vertical

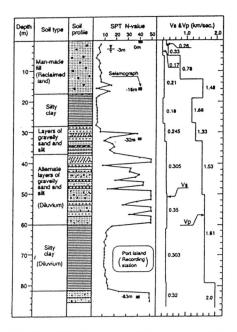

FIG. 1. Soil Profile at the Site of Vertical Array of Seismograph at Port Island (Toki 1995)

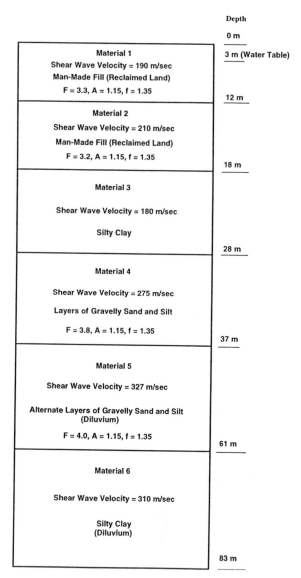

FIG. 2. Representative Shear Wave Velocity Profile and Electrical Parameters of Instrumented Site at Port Island, Kobe

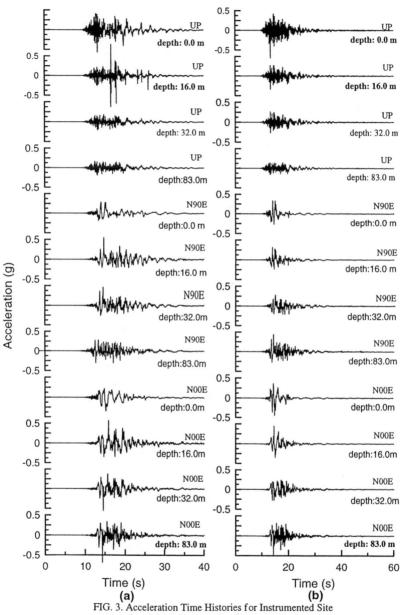

FIG. 3. Acceleration Time Histories for Instrumented Site
at Port Island, Kobe (a) Recorded ; (b) Evaluated

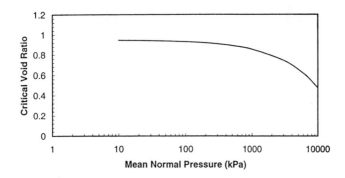

FIG. 4 In-Situ Critical Void Ratio Variation with Mean Normal Pressure for Sands at the Instrumented Site Port Island, Kobe

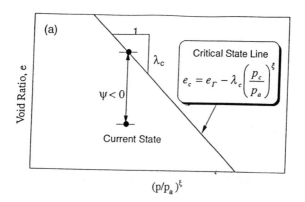

FIG. 5. Void Ratio (e) Variation with $\left(\dfrac{p}{p_a}\right)^{\xi}$

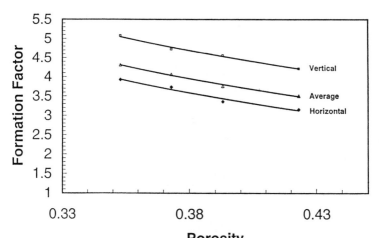

Porosity
FIG. 6. Vertical, Horizontal and Average
Formation Factors of Kobe Sand

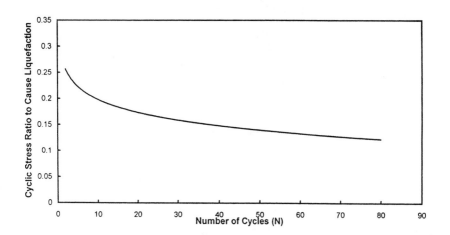

FIG. 7. Relationship Between Cyclic Stress Ratio to Cause
Liquefaction and Number of Cycles of Loading

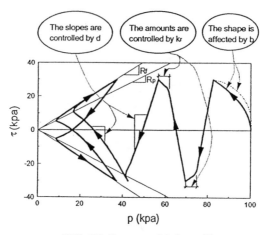

FIG. 8 Influences of d, k_r and b

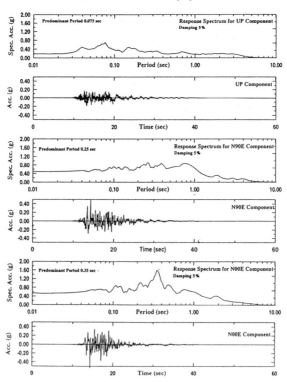

FIG. 9 Acceleration Time Histories and Response Spectra for Port Island Strong Motions at – 83 m (Input Motions)

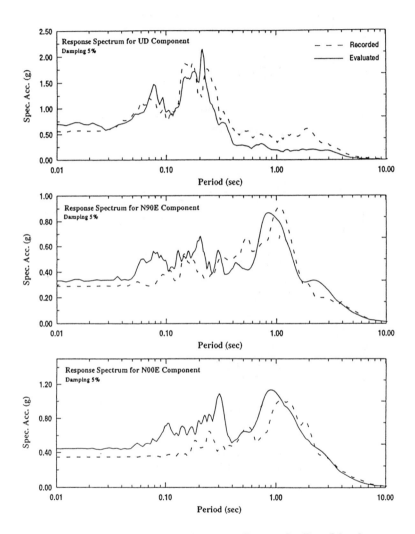

**FIG. 10 Comparison of Response Spectra for Port Island
Strong Motions at Ground Surface**

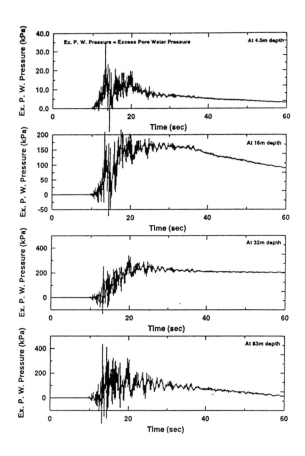

FIG. 11 Evaluated Excess Pore Water Pressure Time Histories
for Instrumented Site at Port Island, Kobe

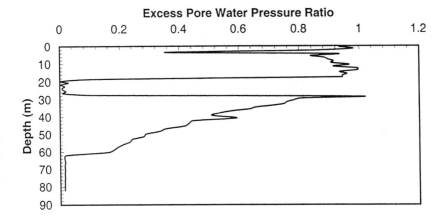

FIG. 12. Maximum Excess Pore Water Pressure Ratio Variation with Depth (Instrumented Site at Port Island, Kobe)

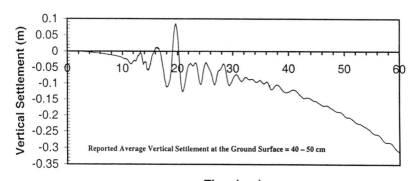

Time (sec)
FIG.13. The Evaluated Vertical Settlement Time History of Instrumented Site at Port Island

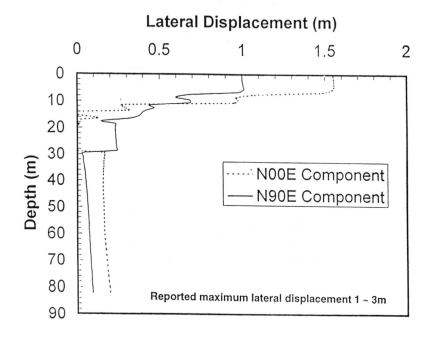

**FIG. 14 The Evaluated Lateral Displacement Variation with Depth at
the end of Shaking (N00E Component and N90E Component)**

Table 1. State and Model Parameters for Instrumented Site at Port Island, Kobe

STATE PARAMETERS:	Material 3	Material 6
Porosity (n)	0.5	0.5
Permeability(k in m/sec)	1.00E-07	1.00E-04
Coefficient of earth pressure at rest(Ko)	0.7	0.7

MODEL PARAMETERS:

Φ (Deg)	30	30
Go	222.9	373.6
λ	0.0421	0.0251
κ	0.00421	0.00251
hr	0.763	0.757
d	100	100
Rp/Rf	1	1
Kr	100	100
b	2	2
hp	35	35

Table 2. State and Model Parameters for Instrumented Site at Port Island, Kobe

STATE PARAMETERS:	Material 1	Material 2	Material 4	Material 5
Porosity (n)	0.424	0.345	0.432	0.42
Permeability(k in m/sec)	5.50E-04	2.30E-04	1.00E-04	1.00E-04
Coefficient of earth pressure at rest(Ko)	0.43	0.44	0.7	0.7

MODEL PARAMETERS:

Φ (Deg)	29.77	29.08	38	38
Go	256.1	232.2	316.6	370.1
κ	0.00215	0.00153	0.00172	0.00147
hr	0.256	0.345	0.243	0.294
d	1.25	1.25	5	5
Kr	0.05	0.05	0.5	0.5
b	2	2	2	2
ζ	0.7	0.7	0.7	0.7
m	3.6	3.6	3.6	3.6
n	0.75	0.75	0.75	0.75
λ_{cr}	0.019	0.019	0.019	0.019
e_r	0.94	0.94	0.94	0.94

EVALUATION OF DYNAMIC SITE RESPONSE AND DEFORMATION OF INSTRUMENTED SITE – ELCENTRO

K. Arulanandan [1] M. ASCE and K. (Siva) Sivathasan [2] S. M. ASCE

ABSTRACT: The observed dynamic response of the instrumented Wildlife site during the 1987 Superstition Hills earthquake was utilized to demonstrate the feasibility of computer simulation of earthquake induced site response and liquefaction-induced deformations of a level ground site. A non-destructive in-situ electrical method was used to obtain the initial state parameters and constitutive model constants representative of the site. The measured shear wave velocity measurements are compared with the shear wave velocity values evaluated using the electrical method. The dynamic response analysis used the effective stress-based, nonlinear, fully coupled, finite element program SUMDES with a reduced order bounding surface hypo-plasticity model to simulate the stress-strain behavior of cohesive soils and modified reduced order bounding surface hypo-plasticity model to simulate the stress-strain behavior of non-cohesive soils. The results of the dynamic analysis showed that the acceleration, excess pore water pressure response histories and liquefaction-induced deformations agreed reasonably well with the observed behavior during the Superstition Hills earthquake. The results of this study show that computer simulation of earthquake effects of a level ground site is possible using non-destructive in-situ testing to obtain the initial state and constitutive model parameters representative of the site and a verified numerical procedure.

INTRODUCTION

Before the event evaluation of the response of our infrastructure systems

---

[1] Professor, Dept. of Civ. and Envir. Engrg., Univ. of California, Davis, CA 95616.
[2] Graduate Student, Dept. of Civ. and Envir. Engrg., Univ. of California, Davis, CA 95616.

subjected to earthquakes is a desirable objective for the adoption of appropriate, safe and economical retrofit measures of existing facilities and the safe and economical design of new infrastructure facilities. Once a verified numerical procedure and in-situ properties representative of the site are available, computer simulation of earthquake effects of any site can be carried out as demonstrated in a recent paper "Numerical Simulation of Liquefaction-Induced Deformations" Arulanandan et al. 2000. The first objective of this paper is to provide an additional example to evaluate the earthquake response of the instrumented site (Wildlife site California) where the pore pressure and acceleration time responses and liquefaction-induced deformation were recorded during the Superstition Hills earthquake 1987. Wildlife site, California is the only instrumented site where acceleration time responses, pore pressure time responses, vertical and lateral deformations were recorded (Holzer et al. 1989a). The second objective of this paper is to demonstrate how the initial state parameters and constitutive model constants representative of the site were obtained using shear wave velocity and in-situ formation factor measurements (ERTEC WESTERN, INC. 1982). In the earlier paper "Numerical Simulation of Liquefaction-Induced Deformations" Arulanandan et al. 2000, the initial state parameters and constitutive model constants were obtained using laboratory formation factor measurements on remolded samples and shear wave velocity measurements.

EARTHQUAKE AND SOIL CHARACTERISTICS AT INSTRUMENTED SITE

On November 24, 1987, at 05:15 PST (Pacific Standard Time) the Superstition Hills, California earthquake struck the instrumented Wildlife site, adjacent to the Alamo river in the Imperial valley, caused liquefaction and developed lateral displacements toward the river ranging from 0 to 230 mm. Extensive ground cracking indicative of lateral spreading accompanied liquefaction at the array. The site is 23 km east of the epicenter of the Elmore ranch earthquake and 31 km east-northeast of the epicenter of the Superstition Hills earthquake. A location map of the liquefaction array and earthquake epicenters, a stratigraphic cross-section of the array, and a schematic of instrument deployment are shown in Figs. 1a and 1b. A representative soil profile of the Wildlife site is shown in Figs. 2a and 2b. The site had been instrumented with both accelerometers and electrical piezometers and, for the first time, excess pore water pressure ratios of 100% were measured in the field in a saturated silty sand site during an earthquake (Youd and Bartlett 1988, Holzer et al. 1989 a, b). Six pore water pressure transducers and two three component force-balance accelerometers, one at the surface and one downhole, were installed at the array. Instrumentation was installed in 1982. Shallow deposits at the array consist of saturated, flood plain sediments that fill an old incised channel of the Alamo River. The uppermost unit at the array is a 2.5 m thick flat lying silt bed that overlies the unit that liquefied a 4.5 m thick silty sand. Beneath these floodplain deposits is a 5.0 m thick silty clay unit, the uppermost unit of a dense and regionally extensive sedimentary deposit. The Alamo River, a perennial stream because of drainage from irrigation, currently occupies a 3.7 m deep channel 23 m east of the center of the array and controls the water table depth at about 1.2 m. The soil properties of the site had been thoroughly measured prior to the earthquake. Therefore, the 1987 Superstition Hills, California earthquake records provide researchers with a unique opportunity to improve the understanding of the mechanics of seismic pore pressure build up and liquefaction, as well as the understanding the complex relation between pore pressure, water flow, cracking and permanent displacements during lateral spreads.

FIELD INVESTIGATIONS

Seismic shear wave velocity tests and in-situ resistivity measurements were performed at the Wildlife site on February 9, 10, 11 and 12, 1982. The field resistivity measurements were obtained using a Geoelectronics Electrical Soil Probe Model GE-100 (see Figs.3a and 3b). The electrical soil probe consists of a 3-inch diameter, 5 foot long tube with one foot hollow section at the end. The one foot hollow section is similar to a conventional thin wall sampler, except that it contains three oriented electrodes located on the surface of the inside wall. Aggregate soil resistivity is measured in two directions between the three electrodes. In the probe's solid section, an electric pump draws fluid from the soil into an electrode cavity where water resistivity measurements are automatically taken.

A drill rig is used to bore the hole to push the probe 9 inches into the undisturbed soil at the borehole bottom. The operator then balances the down-hole resistivity bridge from the surface control box. The bridge impedence is incrementally stepped by digital electronic messages sent from the microprocessor housed in the instrument case or, if necessary, the bridge is balanced manually. Separate readings from the horizontal electrode set, the vertical electrode set, and the water cavity electrode set are then taken.

Crosshole shear wave velocity surveys were conducted at the field site. The surveys were performed to determine the compressional (P–wave) and the shear wave (S–wave) velocities of the subsurface materials. During a crosshole survey, the seismic waves were generated at a specific depth in a boring. The waves were detected by geophones which are positioned at the same depth in adjacent borings. The result of a crosshole survey is a profile of horizontally measured seismic velocities varying with depth. Seismic waves were generated by striking the top of the drill rod in the source boring with a sledge hammer while the drill bit was in contact with the formation. This procedure generated relatively large amplitude, vertically oriented, S – waves and much lower level P – waves in the horizontal travel path. Cross-hole shear wave velocity measurements are compared with the shear wave velocities evaluated using the electrical method (see Fig. 3c).

RESPONSE STUDIES UTILIZING WILDLIFE RECORDS

Several investigators have utilized records from the Wildlife site to analyze site and pore-pressure responses during the 1987 earthquake. Here we only mention studies by Zeghal and Elgamal (1994), Vucetic and Thilakaratne(Department of Civil Engineering, University of California, Los Angeles, 1989), Dobry et al. (1989) and Gu et al. (1994).

Vucetic and Thilakaratne (1989) and Dobry et al. (1989) performed site response analyses using the computer program DESRAMOD. Dobry et al. (1989) reported results of both simplified calculations and site response analyses using the program DESRAMOD. Only excess pore water pressure predictions were reported in their paper.

Zeghal and Elgamal (1994) utilized the recorded surface and downhole (at 7.5 m depth) acceleration , and the excess pore water pressures measured by P5 (at 2.9 m depth), to estimate: (1) Average shear stress-strain history within the top 7.5 m layer of sediments and (2) effective stress and strain paths at the elevation of piezometer P5 (2.9 m depth). A simple procedure was adopted to obtain direct estimates based on these three seismic records exclusively. Seismic soil behavior was analyzed by investigating: (1) The variation of shear stress as a function of average shear strain; (2) the variation of shear stress and average shear strain as a function of effective vertical pressure (at P5, 2.9m depth). As acceleration records are available only at the surface and down-hole stations, linear interpolation was utilized to evaluate shear stress and strain histories within the top 7.5 m layer.

Gu et al. (1994) conducted the stress redistribution analysis under the fully undrained condition to consider the effects of strain-softening behavior of liquefied materials and the re-consolidation analysis using Biot's theory to consider the effects of dissipation of excess-pore water pressures. In the stress redistribution (static) analyses, a simplified undrained boundary-surface model (Gu el al. 1992) was used to simulate the behavior of liquefiable soils and the hyperbolic strain-softening model, introduced by Chan and Morgenstern (1989), was adopted to simulate the soil behavior during collapse from its peak strength to steady-state strength. The collapse surface proposed by Sladen et al. (1985) was embedded in the model as a triggering condition for the onset of strain-softening behavior of liquefied materials. In the re-consolidation analyses, elastic models were used. The same finite-element mesh for stress re-distribution analyses was used for the re-consolidation analyses.

In the above procedure the response of soil is somehow predetermined. Before-liquefaction and post-liquefaction are treated as two separate processes. In such a procedure, liquefaction and post-liquefaction deformations are triggered by certain preset criteria. The sensitivity of the triggering criteria to the analysis and the impact of the uncertainties in the evolution of soil states, which the liquefaction triggering depends on, have not been addressed in detail. Events before and after liquefaction are essentially two divided stages of a single continuous process, and a unified procedure is preferable to treat earthquake response including the post liquefaction deformation. A unified procedure needs a unified constitutive model, in particular a model that can reproduce all significant stress-strain responses of granular soils during the entire process of earthquake motion. In addition, the response of the model should be loading history dependent, i.e., not predetermined by the analysis.

In this paper, a unified, verified, effective stress based, non-linear, fully coupled, finite element method-based procedure is utilized to analyze the instrumented Wildlife site. The constitutive model used in this analysis, can simulate flow liquefaction as well as cyclic mobility. In this model, there is no need to predetermine which soil is flow liquefaction and which is not, no need to have a rather unnatural liquefaction triggering mechanism, and no need to separately define the behavior of the material model for before- and post-liquefaction events.

VERIFIED NUMERICAL PROCEDURE SUMDES

SUMDES is a computer program to perform dynamic response analyses of Sites Under Multi-Directional Earthquake Shaking (Li et al. 1992). This procedure is formulated on the basis of the effective stress principle, vectored motion, transient pore fluid movement, and generalized material stiffness; therefore, it is capable of predicting three-directional motions, the pore pressure buildup, and dissipation within the soil deposits. The overall procedure consists of two computer programs: SUMDES and

TESTMODL. SUMDES performs the site response analyses; TESTMODL is used to examine the responses of the built-in inelastic models with given model parameters and loading conditions. This numerical procedure was verified during the VELACS project. Numerous examples have been published that demonstrate the applicability of the methodology, which utilizes the numerical procedure and the non-destructive in situ electrical method of site characterization (Arulanandan et al. 1982; Arulanandan et al. 1986; Arulanandan et al. 1989; Reyes et al. 1996; Arulanandan et al. 1996; Sivathasan et al. 1998; Arulanandan and Sivathasan 1998, Arulanandan et al. 1999).

CONSTITUTIVE MODELS USED IN THIS ANALYSIS

1. Reduced Order Bounding Surface Hypo-plasticity Model:

The reduced order bounding surface hypo-plasticity model (Li et al. 1992, Li 1996) is a special version of the original bounding surface hypo-plasticity model (Wang et al. 1990). The model has a smaller number of model parameters than the original model (due to the special loading conditions imposed by site response analysis), but still retains the capability to simulate liquefaction potential and pore water pressure generation. In the reduced order bounding surface plasticity model, ten parameters (see Table 1) are involved in the site response analysis. Among the ten, three parameters (b, h_p and R_p/R_f) are either inactive in the given loading conditions or vary within a small range, so they can be assumed empirically. Two parameters d and k_r can be calibrated based on cyclic strength data obtained from field measurement. Other parameters can be obtained from established correlations between electrical parameters and required parameters.

2. Modified Reduced Order Bounding Surface Hypo-plasticity Model:

The reduced order hypo-plasticity model is modified according to a recent paper by Li et al. (1999) to improve the capability in modeling the constitutive behavior at high strains and most importantly, to incorporate the critical state concept in the sand response. The modification consists of rendering the phase transformation line a function of the state parameter ψ, which measures the difference between the current and critical void ratios at the same mean normal pressure, p, such that when the state parameter is zero, the phase transformation line becomes identical to the critical state line in $q - p$ space. This idea was originally proposed and applied in another constitutive model by Manzari and Dafalias (1997). As a result of this modification, the dilatancy depends on the state in a way which yields a zero value at critical state. This dependence allows a realistic modeling of the response of a sand in either loose or dense state, or in the transition from one state to another state. After the modification, dilatancy is not only a function of stress state but also a function of internal material state. In this model, twelve parameters (see Table 2) are involved in the site response analysis. Among the twelve, d parameter can be calibrated based on cyclic strength data obtained from field measurement as described below. Locating the critical state line in the e-p space is of critical importance for the determination of the state parameter, ψ. It has been shown that the critical state line in e-p space can be closely approximated by considering the relationship between the average shape factor and the position of the critical state line of sand based on laboratory test data. For the Wildlife site sand, the location of the critical state line in e-p space is shown in Fig. 4 (Arulanandan 1994; Arulanandan and Sivathasan

1995). Other parameters can be obtained from established correlations between electrical parameters and required parameters (see Arulanandan et al. 2000).

PARAMETER BASED ON CYCLIC STRENGTH DATA (d)

The cyclic strength is estimated based on the in situ electrical method (Arulmoli et al. 1985). For example, cyclic strength curve for electrical parameter $A^3 /(\overline{Ff}) = 0.26$ is shown on Fig. 5. This curve can be simulated by running the utility program TESTMODL from the program package of SUMDES for a single element response with other calibrated parameters controlling the development of pore pressure under undrained conditions:

The calibration procedure to find d is a process of trial and error:

1. Select cyclic stress ratio, with a specified mean normal confining stress;

2. Assume a value of d;

3. Run the utility program TESTMODL for the selected stress level and count the number of cycles to cause liquefaction (see Fig 6);

4. Compare the results obtained from step 3 against the given relationship of stress ratio required to cause liquefaction versus number of cycles to cause liquefaction. Adjust the d value if needed;

5. Repeat steps 3 and 4 until satisfactory result is obtained.

Repeat the same procedure for two or more shear stress amplitudes, with a specified mean normal stress.

NON-DESTRUCTIVE ELECTRICAL METHOD OF IN-SITU CHARACTERIZATION

Considerable research, for example the VELACS project, has been undertaken to verify numerical procedures using laboratory test results and known initial state parameters. Difficulty in obtaining undisturbed samples and carrying out reliable laboratory tests to avoid disturbance of the sample to calibrate constitutive models seem to be a major obstacle in advancing the use of fully coupled, effective stress-based, nonlinear, numerical procedures using elastoplastic constitutive models. The in-situ non-destructive site characterization is necessary to get the initial state parameters and constitutive model constants representative of the site for the computer simulation of the site to earthquakes.

It has been shown experimentally by numerous studies conducted by Archie (1942), Arulanandan and Kutter (1978) and Wyllie and Gregory (1957) that a non-dimensional electrical parameter, called the formation factor, is dependent upon the *particle shape, long axis orientation, contact orientation* and *size distribution*, and also *cementation, degree of saturation, void ratio* and *anisotropy*. The formation factor is thus a function of the fabric of the soil and varies with stress to the extent that the fabric is altered by the stress. An index, the formation factor F, is experimentally obtained as the ratio of the conductivity of the pore fluid and the conductivity of the soil sample. Thus, the formation factor is a non-dimensional parameter. It depends on sand structure, especially on its elements associated with particle arrangement , particle shape, and porosity. The dependence of the formation factor on void ratio, particle shape and long axis orientation has been shown theoretically by Arulanandan and Dafalias (1979), and Dafalias and Arulanandan (1979). The formation factor F has been shown to be a valuable directional index parameter when the sand structure is anisotropic. Due to the above facts, in-situ measurements of formation factor (F) provides a means to characterize the structure of sands in the field and to obtain empirical correlations between shape and the anisotropic state of sand characterized in terms of F and the engineering behavior such as liquefaction characteristics of sands.

The formation factor can be used to quantify and predict the porosity, shape and anisotropy of sand deposits. An average formation factor $\overline{F}$, is given by

$$\overline{F} = \frac{(F_v + 2F_h)}{3} \tag{1}$$

where, F_v is the vertical formation factor and F_h is the horizontal formation factor.

It has been shown that $\overline{F}$ is independent of anisotropy caused by the orientation of preferred particles, and is a direct measure of porosity (n), for a given sand. For practical purposes, an integration technique proposed by Bruggeman (1935) was used by Dafalias and Arulanandan (1979) to derive an expression for the average formation factor, $\overline{F}$, as a function of porosity (n), and average shape factor ($\overline{f}$), as

$$\overline{F} = n^{-\overline{f}} \tag{2}$$

It has been shown both theoretically and experimentally that the shape factor is directional and depends on porosity, gradation, the particles' shape and orientation of the particles (Dafalias and Arulanandan 1979, Arulanandan and Kutter 1978). An anisotropy index (A), was introduced by Arulanandan and Kutter (1978) as

$$A = \left(\frac{F_v}{F_h} \right)^{0.5} \tag{3}$$

The anisotropy of sand structure is due to the orientation of individual particles and the contact orientation.

METHOD OF ANALYSIS

Samples were taken from the silty sand layer at the instrumented site at wildlife site, California, and a correlation between the formation factor and porosity was established (see Fig. 7). From the non-destructive electrical measurements, the average shape factor and anisotropy index were calculated. The average formation factor was calculated using in-situ non-destructive resistivity measurements in two different directions. Porosity was obtained from the pre-established relationship (see Fig. 7) between formation factor and porosity using the calculated average formation factor. All the constitutive model constants and initial state parameters were calculated using the average formation factor, average shape factor and anisotropy index (see Arulanandan et al. 2000).

COMPARISON OF SHEAR WAVE VELOCITY MEASUREMENTS

A correlation between K_{2max} and the electrical parameter $\bar{F}/(A\bar{f})^{0.5}$ is given by Arulanandan et al. (1983) and is used to obtain K_{2max}. K_{2max} depends on void ratio, strain amplitude, geological age of the sand mass and in situ stresses. The mean effective confining pressure, P, can be calculated using site configuration and unit weight of the soil mass. The maximum shear modulus can be obtained using

$$G_{max} = 1000K_{2max}(P)^{0.5} \qquad (4)$$

where P is given in pounds per square foot and G_{max} is given in pound per square foot. Using known maximum shear modulus and density of soil mass, one can obtain the shear wave velocity

$$V_s = \sqrt{\frac{G_{max}}{\rho}} \qquad (5)$$

The comparison of the shear wave velocity values calculated using above procedure with the measured cross hole shear wave velocity values as shown in Fig. 3c demonstrate a reasonable agreement.

ANALYSIS RESULTS

The numerical simulation of the Wildlife instrumented site, California was performed by the computer program SUMDES. In the analysis, multi-directional shaking was applied to the soil deposits. The input motions and their response spectrum are shown in Fig. 8. The reduced order bounding surface hypo-plasticity model was used to simulate the cohesive soil behavior and the modified reduced order bounding surface hypo-plasticity model was used to simulate the non-cohesive soil behavior. Calibrated constitutive model constants and state parameters are provided in Tables 1 and 2.

From the specification for the instrumented array at Wildlife site, California, accelerometers were placed at two different depths. At each depth three different directional components of acceleration were reported. In the present analysis, all the predictions and comparisons were made at the two different depths in three different directions.

Recorded and evaluated accelerations at different depths in different directions are presented in Figs. 9a and 9b. Response spectra from analysis results and recorded accelerations at the ground surface are compared in Fig. 10. Note that recorded and SUMDES results compare favorably. Recorded and evaluated excess pore water pressure at different depths are shown in Figs. 11a and 11b, respectively. Field pore water pressure measurements were reported for comparison with the evaluated pore water pressure during the earthquake shaking at four different depths. The maximum pore water pressure ratio variation with depth is shown in Fig. 12. It shows that the zone of high pore water pressure ratio in excess of 90% extended in the first 6.8m below water table. Pore pressure time histories, in terms of pore-pressure ratio, recorded and evaluated are shown in Figs. 13a and 13b, respectively. Vertical settlement variation with time at the ground surface is shown in Fig. 14. Reported average vertical settlement at the ground surface is about 5 cm. Evaluated lateral displacement variation with depth is shown in Fig. 15. Reported maximum lateral displacement 230 mm occurred at the waterfront (Dobry et al. 1989). The observed lateral displacements were generally large near the waterfront and decreased with distance inland. These results provide reliability to the computational analysis.

CONCLUSIONS

The pore pressure, accelaration and liquefaction-induced deformation time histories were simulated using the nonlinear, fully coupled, effective stress based method of analysis. The evaluated acceleration and the excess pore pressure time histories are in close agreement with the recorded acceleration and the excess pore pressure time histories at the Wildlife site, respectively . This analysis shows that liquefaction occurred in the silty sand layer and agrees with observed liquefaction phenomenon at the Wildlife site, California. Evaluated vertical settlement and lateral displacements are in close agreement with the observed behaviors at the Wildlife site.

It has been demonstrated that it is possible to evaluate the dynamic response of a level ground instrumented site using verified numerical procedure and input properties such as initial state parameters (porosity, coefficient of earth pressure at rest, hydraulic conductivity) and constitutive model constants (slope of the rebound line, friction angle, maximum shear modulus, slope of the critical state line in e_c - p space, hardening parameters, etc.) representative of the site obtained by non-destructive in situ testing and shear wave velocity measurements. The comparison of the cross hole shear wave velocity values obtained by wave propagation method with the shear wave velocities obtained by the electrical method are shown to be in close agreement.

ACKNOWLEDGMENT

The authors are indebted to Dr. X. S. Li and Prof. Y.F. Dafalias for reviewing the manuscript and providing valuable comments.

Appendix I. References

Archie, G. E. (1942). "The Electrical Resistivity Log as an Aid in Determining Some Reservoir Characteristics," Transactions, American Institute of Mining, Metallurgical and Petroleum Engineers, Vol. 146, pp.54-61.

Arulanandan, K. (1994). " In-Situ Soil Characterization, Location of Critical Void Ratio Variation with Mean Normal Stress by In-Situ Testing," UC Davis Report and lecture presented at the Kyoto University, Kyoto, Japan.

Arulanandan, K., Anandarajah, A., and Meegoda, N. J. (1983). "Soil Characterization for Non-Destructive In-Situ Testing," Symposium Proceedings part 2, The interaction of Non-Nuclear Munitions with Structures, US Air Force Academy, Colorado, May 10 – 13.

Arulanandan, K., Arulmoli, K. and Dafalias, Y. F. (1982). "In Situ Prediction of Dynamic Pore Pressures in Sand Deposits," International Symposium on Numerical Models in Geomechnics, Zurich, Switzerland, 13-17 September, pp. 359-367, A. A. Balkema/Rotterdam.

Arulanandan , K., and Dafalias, Y. F. (1979). "Significance of Formation Factor in Sand Structure Characterization," Letters in application and Engineering Sciences, Vol. 17, pp.109-112.

Arulanandan, K. and Kutter, B. L. (1978). "Directional Structure Index Related to Sand Liquefaction," Proceedings of the June 19-21, Specialty Conference on Earthquake Engineering and Soil Dynamics, ASCE, Pasadena, California, pp. 213-229.

Arulanandan, K., Li, X. S., Paulino, G. H. and Sivathasan, K. (1996). "Dynamic Response of Saturated Level Ground Sites using Verified Numerical Procedure and In-Situ Testing," Proceedings of the National Science Foundation and California

Transportation Dept. Sponsored Workshop/Conf. on Application of Numerical Procedures in Geotech. Earthquake Engrg., Univ. of California, Davis.

Arulanandan, K., Li, X. S. and Sivathasan, K. (2000). "Evaluation of Dynamic Site Response and Deformation of Instrumented Site-Kobe," Journal of Geotechnical and Geoenvironmental Engineering, ASCE, (in press).

Arulanandan, K., Muraleetharan, K. K., Dafalias, Y. F., Shinde, S. B., Kaliakin, V. N. and Herrmann, L. R. (1989). "Pore Pressure and Lateral Stresses Using In Situ Properties," Proceedings of the Twelfth International Conference on Soil Mechanics and Foundation Engineering, Rio De Janerio, 13-18 August, A. A. Balkem/Rotterdam/Brookfield.

Arulanandan, K. and Sivathasan, K. (1995). "In-Situ Prediction of Critical Void Ratio Vs Mean Normal Pressure," Technical Report, University of California, Davis.

Arulanandan, K. and Sivathasan, K. (1998). "Evaluation of Site Response and Deformation of Instrumented Bridge Sites Subjected to Large Magnitude Earthquakes," Preliminary Report to Dept. of Transportation, California.

Arulanandan, K., Yogachandran, C., Meegoda, N. J., Liu Ying and Shi Zhauji (1986). "Comparison of the SPT, CPT, SV and Electrical Methods of Evaluating Earthquake," Proceedings of In Situ '86, GT Div., ASCE, June 23-25, Blacksburg, VA.

Arulmoli, K., Arulanandan, K. and Seed, H. B. (1985). "New Method for Evaluating Liquefaction Potential," Journal Geotechnical Engineering Division, ASCE, 111(1), 95-114.

Bennett, M. J., McLaughlin, P. V., Sarmieto, John and Youd, T. L. (1984). "Geotechnical Investigation of Liquefaction Sites, Imperial Valley, California," Open File Report 84-252, U. S. Geological Survey, Denever, Colorado, 1-103.

Bruggeman, D. A. G. (1935). "Berechung Verschiedenez Physikalischer Konstanten Von Heterogenen Substanzen," Ann. Phys. Lpz. 5, Vol. 24, pp. 636.

Chang, C. S. and Misra, A. (1990). "Packing Structure and Mechanical Properties of Granulates," J. of Engineering Mechanics, ASCE, Vol. 116, No. 5, pp. 1077-1093.

Dafalias, Y. F. and Arulanandan, K. (1979). "Electrical Characterization of Transversely Isotropic Sands," Archives Of Mechanics, Warsaw, 31(5), 723-739.

Dobry, R., Elgamal, A. W., Baziar, M. and Vucetic, M. (1989). "Pore Pressure and Acceleration Response of Wildlife Site During The 1987 Earthquake,"

ERTEC WESTERN, INC. (1982). "Investigation of: Correlation Between Electric Resistivity Measurements and Crosshole Shear Wave Velocity Measurements," Preliminary Report, The Earth Technology Corporation, 3777 Long Beach Boulevard, Long Beach, CA 90807. December, Ertec I. D. 82-236-(1-10)

Gu, W. H., Morgenstern, N. R. and Robertson, P. K. (1992). "Progressive Failure of Lower San Fernando Dam," J. Geotech. Engrg., ASCE, 119(2), 333-348.

Gu, W. H., Morgenstern, N. R. and Robertson, P. K. (1994). "Postearthquake Deformation Analysis of Wildlife Site," J. Geotech. Engrg., ASCE, Vol. 120, No. 2, February.

Holzer, T. L., Youd, T. L. and Hank, T. C. (1989 a). "Dynamic of Liquefaction During the 1987 Superstition Hills, California Earthquake," Science, 244, pp. 1-116.

Holzer, T. L., Youd, T. L. and Bennett, M. J. (1989 b). "In Situ Measurements of Pore Pressure Build-Up During Liquefaction," Proceedings of the 20th Joint Meeting of the U. S. – Japan Cooperative Program in Natural Resources Panel on Wind and Seismic Effects, Published by the National Institute of Standards and Technology, Gaithersburg, MD 20899.

Li, X. S. (1996)." Reduced-Order Sand Model for Ground Response Analysis," Journal of Engineering Mechanics, ASCE, 122(9), pp. 872-881.

Li, X. S., Dafalias, Y. F. and Wang, Z. L. (1999). "A Critical-State Hypo-plasticity Sand Model with State Dependent Dilatancy," Canadian Geotechnical Journal (accepted).

Li, X. S., Wang, Z. L., and Shen, C. K. (1992). "SUMDES: A Non Linear Procedure for Response Analysis of Horizontal-Layered Sites Subjected to Multi-Directional Earthquake Loading," Department of Civil and Environmental Engineering, University of California at Davis.

Manzari, M. T. and Dafalias, Y. F. (1997). "A Critical State Two-Surface Plasticity Model for Sands," Geotechnique, Vol. 47, No. 2, pp. 255-272.

Reyes, C., Arulanandan, K., Steve Mahnke, Chad Baker and Sivathasan, K. (1996). "Fully Coupled Effective Stress Based Analysis to Investigate the Consequences of Soil Liquefaction at Mosher Slough for FEMA Project," Report to Kleinfelder and Associates, Stockton, California.

Sivathasan, K., Paulino, G. H., Li, X. S. and Arulanandan, K. (1998). "Validation of Site Characterization Method for the Study of Dynamic Pore Pressure Response," Geotechnical Special Publication No. 75, Volume one, Geotechnical Earthquake Engineering and Soil Dynamics III, ASCE, Seattle, Washington.

Sladen, J. A., D'Hollander, R. D. and Krahn, J. (1985) "The Liquefaction of Sand, a Collapse Surface Approach," Canadian Geotechnical Journal, 22(4):564-578.

Vucetic, M. and Thilakaratne, V. (1989). "Liquefaction at the Wildlife Site-Effect of Soil Stiffness in Seismic Responses," Proceedings of the 4th International Conference on Soil Dynamics and Earthquake Engineering, Mexico City, Mexico.

Wang, Z.L., Dafalias, Y.F. and Shen, C.K. (1990). "Bounding Surface Hypo-plasticity Model for Sand," Journal of Engineering Mechanics, ASCE, Vol.116, No. 5, May, pp983-1001.

Wyllie, M. R. J. and Gregory, A. R. (1953). "Formation Factors of Unconsolidated Porous Media," Influence of Particle Shape and Effect of Cementation, Petroleum Transactions, American Institute of Mining, Metallurgical and Petroleum Engineers, Vol. 198, pp.103-109.

Youd, T. L. and Bartlett, S. F. (1988). " US Case Histories of Liquefaction-Induced Ground Displacement," First US-Japan Workshop on Liquefaction, Large Ground Deformation and their Effect on Lifeline Facilities, Tokyo, Japan.

Zeghal, M. and Ahmed-W. Elgamal (1994). "Analysi of Site Liquefaction Using Earthquake Records," Journal of Geotechnical Engineering, ASCE, Vol. 120, No. 6, June.

Appendix II. Notation

The following symbols are used in this paper:

A	= anisotropy index
b	= parameter affecting the shape of the stress paths of the virgin shear loading
d	= parameter which characterizes the rate of the effective mean normal stress change caused by shear unloading
$\bar{F}$	= average formation factor
F_h	= horizontal formation factor
F_v	= vertical formation factor
$\bar{f}$	= average shape factor
G_{max}	= maximum shear modulus
G_o	= constant related to shear modulus
h_p	= parameter controlling the amount of shear strain increment due to the change of the maximum effective mean normal stress in the loading history
h_r	= parameter which characterizes the relationship between shear modulus and shear strain magnitude
K_o	= coefficient of earth pressure at rest
K_{2max}	= modulus related to maximum shear modulus
k	= coefficient of hydraulic conductivity
k_r	= parameter which characterizes the amount of the effective mean normal stress change caused by shear loading

m   = model constant

n   = model constant

n   = porosity

q   = deviatoric stress

R_p/R_f = ratio of stress ratio at phase transformation state and stress ratio at failure

          state

ϕ = friction angle

κ = slope of the rebound line in e-ln(p) space

ξ = constant

ψ = state parameter

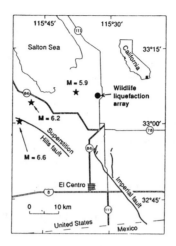

Fig. 1a Location Map of Liquefaction Array and Earthquake Epicenters (Holzer et al. 1989)

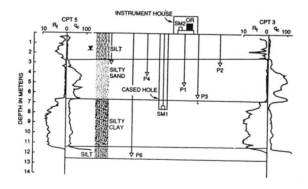

Fig. 1b Stratigraphic Cross Section of Array and Schematic of Instrument Deployment (Bennett et al. 1984)

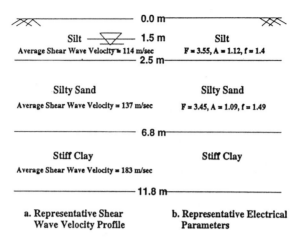

a. Representative Shear
Wave Velocity Profile

b. Representative Electrical
Parameters

Fig. 2 Representative Soil Profile of Wildlife Site

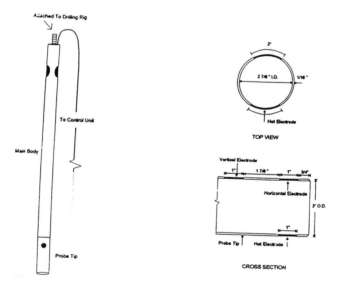

Fig. 3a Electrical Field Probe

Fig. 3b Schematic View of Electrodes
(I. D. – Inner Dia.   O. D. – Outer Dia.)

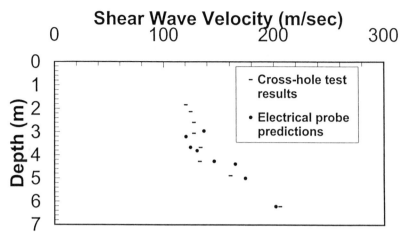

**Fig. 3c Shear Wave Velocity Profile
(Wildlife Site, California)**

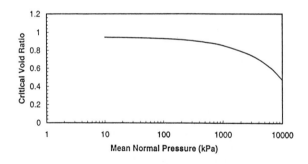

Fig. 4 In-Situ Critical Void Ratio Variation with Mean Normal Pressure for Silty
Sand at the Instrumented Wildlife Site

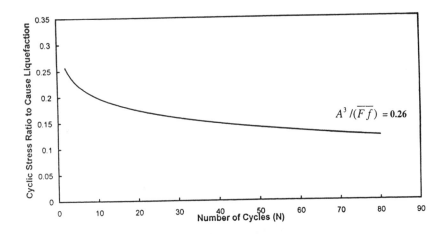

Fig. 5 Relationship Between Cyclic Stress Ratio to Cause Liquefaction and Number of Cycles of Loading

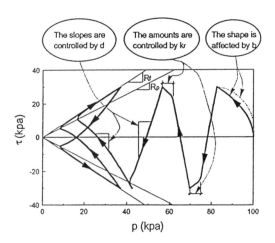

Fig. 6 Influences of d, k_r and b

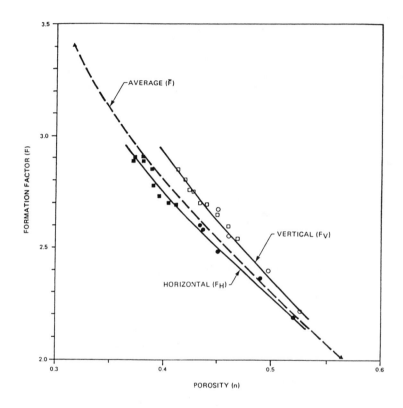

**Fig. 7 Vertical, Horizontal and Average Formation Factors of Wildlife Site Silty Sand
(Ertec Western Inc. 1982)**

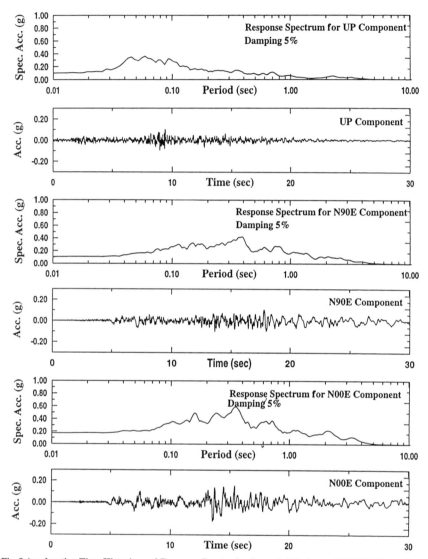

Fig. 8 Acceleration Time Histories and Response Spectra for Recorded Motions at Wildlife Site at 7.5 m

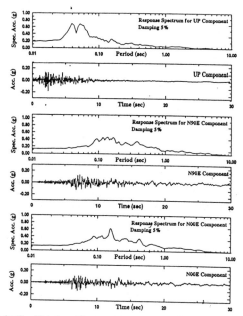

Fig. 9a Acceleration Time Histories and Response Spectra for Motions Recorded at Wildlife Site at 0.0 m

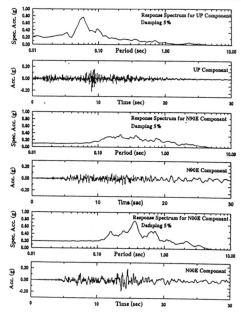

Fig. 9b Acceleration Time Histories and Response Spectra for Motions Evaluated at Wildlife Site at 0.0 m

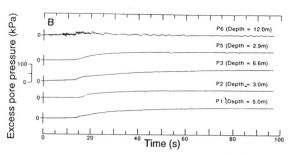

**Fig. 11a Recorded Excess Pore Water Pressure Time Histories
for the Instrumented Wildlife Site (Holzer et al. 1989)**

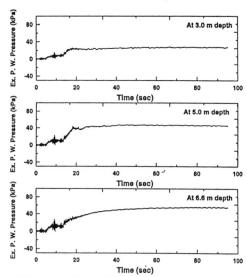

**Fig. 11b Evaluated Excess Pore Water Pressure Time Histories
for the Instrumented Wildlife Site**

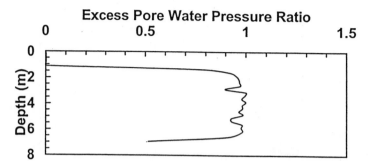

**Fig. 12 Evaluated Maximum Excess Pore
Water Pressure Ratio Variation with Depth
(Wildlife Site, California)**

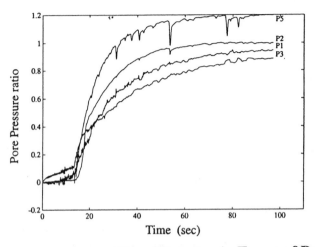

Fig. 13a Pore Pressure Time Histories, in Terms of Pore-Pressure Ratio, Recorded (Wildlife Site, California, Dobry et al. 1989)

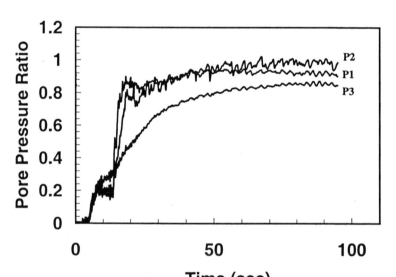

Fig. 13b Pore-Pressure Time Histories, in Terms of Pore-Pressure Ratio, Evaluated (Wildlife Site, California)

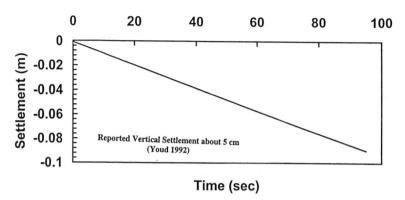

Fig. 14 Evaluated Vertical Settlement
Time History (Wildlife Site, California)

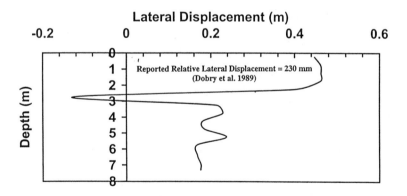

Fig. 15 Evaluated Lateral Displacement
Variation with Depth (Wildlife Site)

Table 1. State and Model Parameters for Cohesive Soil at Instrumented Wildlife Site, California

STATE PARAMETERS:	Material 3
Porosity (n)	0.528
Permeability (k in m/sec)	3.10E-09
Coefficient of earth pressure at rest (Ko)	0.5

MODEL PARAMETERS:

Φ (Deg)	29
Go	529
λ	0.015
κ	0.003
hr	0.705
d	100
Rp/Rf	1
Kr	100
b	2
hp	35

Table 2. State and Model Parameters for Non-Cohesive Soils at Instrumented Wildlife Site, California

STATE PARAMETERS:	Material 1	Material 2
Porosity (n)	0.47	0.44
Permeability (k in m/sec)	3.00E-06	2.10E-05
Coefficient of earth pressure at rest (Ko)	0.58	0.66

MODEL PARAMETERS:

Φ (Deg)	30	32
Go	328	258
κ	0.0028	0.003
hr	1.2	0.46
d	1.3	1.1
Kr	0.3	0.1
b	2	2
ζ	0.7	0.7
m	3.6	3.6
n	0.75	0.75
λ_{cr}	0.018	0.019
e_r	0.98	0.93

Subject Index

Page number refers to the first page of paper

Author Index

Page number refers to the first page of paper